DEVELOPMENT OF AN EFFICIENT MODELLING APPROACH TO SUPPORT ECONOMICALLY AND SOCIALLY ACCEPTABLE FLOOD RISK REDUCTION IN COASTAL CITIES: CAN THO CITY, MEKONG DELTA, VIETNAM

T0262599

Hieu Quang Ngo

DEVELOPMENT OF AN EFFICIENT MODELLING APPROACH TO SUPPORT ECONOMICALLY AND SOCIALLY ACCEPTABLE FLOOD RISK REDUCTION IN COASTAL CITIES: CAN THO CITY, MEKONG DELTA, VIETNAM

DISSERTATION

Submitted in fulfillment of the requirements of
the Board for Doctorates of Delft University of Technology
and
of the Academic Board of the IHE Delft
Institute for Water Education
for
the Degree of DOCTOR
to be defended in public on
Monday, 22 November 2021, at 12:30 hours
in Delft, the Netherlands

by

Hieu Quang NGO
Master of Hydraulic Engineering, Water Resources University
born in Ha Noi, Vietnam

This dissertation has been approved by the
promotors:
Prof.dr.ir. C. Zevenbergen
Prof.dr. R.W.M.R.J.B. Ranasinghe

Composition of the doctoral committee:

Rector Magnificus TU Delft	Chairman
Rector IHE Delft	Vice-Chairman
Prof.dr.ir. C. Zevenbergen	IHE-Delft / TUDelft, promotor
Prof.dr. R.W.M.R.J.B. Ranasinghe	IHE-Delft / University of Twente, promotor

Other member:
Dr. P.D.A. Pathirana   UNDP

Independent members:

Prof.dr.ir. S.N. Jonkman	TU Delft
Prof.dr.ir. A.E. Mynett	TU Delft / IHE Delft
Prof.dr. S.J.M.H. Hulscher	University of Twente
Dr. H.T. Lan Huong	Inst. M.H.C.C. Hanoi Vietnam
Prof.dr. D.P. Solomatine	TU Delft / IHE Delft, reserve member

This research was conducted under the auspices of the Graduate School for Socio-Economic and Natural Sciences of the Environment (SENSE)

CRC Press/Balkema is an imprint of the Taylor & Francis Group, an informa business

Published by:
CRC Press/Balkema
enquiries@taylorandfrancis.com
www.crcpress.com – www.taylorandfrancis.com
ISBN 978-1-0322-2914-0

Acknowledgements

My PhD journey is finally over, and this is my thesis. Completing this PhD was a challenge but also a very interesting and valuable journey for me. Now looking back, I enjoy and cherish every moment of it, including the ups and downs. To bring it to this state, I have received a lot of help, guidance and support from many people throughout this journey and which I would like to take this opportunity to express my most sincere gratitude.

First and foremost, I would like to express my deepest sincere gratitude to my promotor, Professor. Rosh Ranasinghe. Professor, thank you very much for giving me the opportunity to pursue this exciting PhD research under your supervision. It is truly a great honour and pleasure for me, and my heartfelt appreciation for your patience, encouragement, great support and help, also valuable advice about research as well as life's matters. Your dedicated guidance, critical comments, strict requirements with a high standard for quality, depth of knowledge, and innovative insights from the beginning till the end, played a vital role in the success of this dissertation. Also, thank you very much for motivating me to improve both myself and my work for the better. Honestly, I like the way you'd motivated me, even though it was not always "sweet as candy" but it worked for me, and I really appreciate and enjoy that, you know what I meant :).

It is also my great blessing to work under the supervision of Dr. Assela Pathirana. I truly thank you very much for your great guidance and support throughout this journey. I've learned a lot from you and am really inspired by your smart and unconventional ideas to tackle complex problems. I really appreciate your help with not only scientific content but also technical issues that you need a little time to show me how to solve instead of spending a lot much time and effort from my side. Assela, thank you very much also for your patience, encouragement and great advice, which helped me during my PhD research.

I am really grateful to my promotor, Professor. Chris Zevenbergen for his kind support, great help and guidance throughout this process. I really enjoy and appreciate our fruitful discussions. Many thanks for your extensive knowledge and valuable comments, they help me to improve and complete this research.

My special thank to all committee members for evaluating my thesis and for their valuable comments that helped broaden my view from different perspectives.

I thank Professor. Jeroen Rijke, Dr. Juliette Cortes, Dr. Walaa El-Hamamy, Dr. Ngo Anh Quan, Dr. Nguyen Hong Quan, Dr. Tran Dung Duc, Thaine, Dr. Juan Carlos Chacón-Hurtado, Shahnoor, Aya Mohanna, anh Duoc, Chuot, Dr. Vu Quoc Thanh for providing evaluation, and constructive comments and suggestions on improving the *Inform* tool.

I would like to acknowledge the IHE Delft projects OPTIRISK, DURA FR and AXA CC&CR Research fund for providing the funding for this study. Also, I would like to thank the Mekong River Commission for providing 1D ISIS model for the entire Mekong Delta, CH2M Company for delivering the Flood Modeller Pro license, CHI (Computational Hydraulics International) for providing the PCSWMM license and SURFsara for giving the grant to use the e-infra/SURFsara HPC Cloud.

I thank IHE Delft for hosting me during my PhD, and for creating a multicultural and dynamic environment that gives me the opportunity to meet and work with many people from all over the world. Jolanda, Anique, Floor, and Niamh, thank you all very much for the wonderful administrative support during my stay at IHE. Many thanks to the Flood resilience chair group for providing valuable flood resilience knowledge. I have great joy working with colleagues William, Mohan, Dikman, Polpat, Maria, Ha, Nguyen. Mohan, thank you very much for your kindness, great support, and for always being available when I need help. Also, thanks for your constructive comments and suggestions for our publications, I have learned a lot from you during our discussions. Dikman, thank you very much for sharing experience and help. I still remember our coffee breaks, and lunches every Thursday with many funny stories. Those are the pleasant and memorable memories to me. Also, my appreciation goes to all my colleagues and friends in IHE Delft. Jeewa, Janaka, chi Lan, Aries, Aknan, Kelly, Mohaned, Aftab, Ataul, Adele, Milk, thank you all for your kindness, sharing experience, support in many different ways, as well as for all the nice memories that we have shared in Delft.

Special thanks go to my friends here and in my homeland. Thanks for the encouragement, sharing experience in research and life, the distraction, many good laugh and great memories we have shared, which added joy and helped balance my work and life.

Finally, I would like to express my deepest gratitude and love to my family, my parents, and my parents-in-law, relatives for their never-ending care, encouragement, support, and unconditional love. Thank you very much for believing me and supporting me to pursue my own ambitions. Especially, to my beloved wife and my little son (Gau-con), heartily thanks for the love, encouragement, for always being by my side, accompanying me through the ups and downs. You are my strength, my motivation to overcome all difficulties and challenges.

Summary

Coastal cities are among the most urbanized and populated places of the world and are disproportionately exposed to natural disasters including flooding. Flooding in these cities can cause severe impacts on human activities and damage to properties in residential areas. Sea level rise, heavy rainfall, and storm surge drive flooding in these areas and are connected to climate change. Human activities like urbanization, dam construction often exacerbate flood hazard due to amongst other increased runoff due to land-use change and land subsidence.

Assessing flood risk is a critical process for responding to future changes (for example climate change, economic change and population growth) by flood risk management (FRM). FRM in the modern context demands statistically robust approaches due to the need to deal with uncertainties. Quantitative FRM therefore should be addressed probabilistically. However, probabilistic estimates often involve ensemble 2D hydraulic model runs resulting in large computational cost.

Modern FRM in the context of a changing environment, necessitates the involvement of a broad range of stakeholders, who are not necessarily experts in FRM. To be able to implement effective and socially-accepted flood risk reduction measures, it is necessary to have a wider-stakeholder engagement with co-learning and co-designing. This makes it necessary for the flood models, at least at a simplified level, to be understood by and accessible to non-specialists.

To meet both of the above challenges, a flood modelling system is needed that can provide rapid and sufficiently accurate estimates flood risk within a methodology that is accessible to a wider range of stakeholders. This study presents and demonstrates a flood modelling system that contributes towards achieving this goal.

Starting from a detailed one-dimensional model of the Mekong Delta, which was implemented in the accessible and open-source modelling system SWMM of US-EPA, an iterative model simplification process was carried out to obtain fast and accurate water level predictions at the river location near Can Tho – the largest city in the delta. Can Tho was selected as case study throughout this dissertation. The final model obtained was calibrated and validated with observed river flow and water level data for several recorded flood events. The results show that it is possible to simulate river water levels, with an acceptable level of accuracy, at a location of interest in a complex, deltaic river system like the lower Mekong with a relatively simple (simplified from thousands to several tens of cross-sections) and fast (reduction of simulation time from 1.5 h to around one minute, for a 1-year simulation) numerical model.

The simplified 1D model obtained was used to derive the river water levels at Can Tho city under different climatic and hydrologic conditions. The model response to a future

(2050) situation taking into account the impact of climate change (RCP 4.5 and RCP 8.5) on river flow and sea level (tide, sea-level rise, storm surge) was compared with a simulation under current conditions (baseline). For both time periods, flood frequency analyses were performed to determine the river water levels corresponding to a range of return periods. Extreme river water level values were extracted from the results as a set of 24-h long time series around the peak value of each extreme value. These extreme time series were examined for common patterns and two distinct hydrograph patterns (i.e. groups of similar extreme water level time series shapes) were identified. The response of inundation levels in Can Tho to both patterns were then quantitatively compared. The hydrograph pattern with significant influence on flooding was used to drive a 1D/2D coupled model to simulate flood inundation in Can Tho city. Being able to use a representative flood hydrograph pattern in the flood simulations greatly reduced the total number of time-consuming 1D/2D simulations required to obtain probabilistic results. Subsequently, regression relationships between the water level in the river and the maximum inundation level in each computational cell in the 2D model were derived from the 1D/2D flood model results. The maximum inundation depths at all grid cells were then aggregated to generate flood hazard maps for each return period, for both the present and 2050, with and without land subsidence effects.

Results show that if the present rate of land subsidence continues, by 2050 (under both RCP 4.5 and RCP 8.5 climate scenarios), the 0.5 year and 100 year return period flood extents will increase by around 15-fold and 8-fold, respectively, relative to the present-day values. However, without land subsidence, the projected increases in the 0.5 year and 100 year return period flood extents by 2050 (under RCP 4.5 and RCP 8.5) are limited to between a doubling to tripling of the present-day flood extent.

Flood hazard maps obtained were then combined with information on land use for the current and future situations for Can Tho using stage-damage functions to calculate the expected annual damage (EAD). The EAD is projected to increase by 1.7 times and 1.8 times in 2050 due to climate change under both scenarios RCP 4.5 and RCP 8.5 and significantly increase to approximate 20-fold and 21-fold due to the combined effect of climate change and land subsidence.

The simplified 1D model for the entire Mekong Delta, flood hazard and damage maps, as well as the estimated damage obtained directly from the 1D/2D coupled model, were combined to develop an interactive, web-based tool (*Inform*) aimed at supporting a rapid flood risk assessment for co-designing by stakeholders. Limited pilot testing indicated that *Inform* contains the must-have features of a co-design tool (e.g. in-built input library, flexible options, easy to use, user-friendly interface). Moreover, it takes about one minute for *Inform* to compute a water level time series containing the maximum river water level at Can Tho and to generate the resulting flood hazard and damage maps, and estimated damage for a flood event. This fast computation contributes to the user-friendliness of *Inform*.

Samenvatting

Kuststeden behoren tot een van de meest dichtbevolkte plaatsen ter wereld en zijn mede hierdoor buitengewoon kwetsbaar voor de gevolgen van natuurrampen zoals overstromingen. Wanneer deze steden getroffen worden door overstromingen kan dit zeer grote sociale gevolgen hebben en leiden tot grote schade aan de bebouwde omgeving. Zeespiegelstijging, hevige neerslag, en stormen kunnen in deze gebieden overstromingen veroorzaken en klimaatverandering heeft hier invloed op. Menselijke activiteiten, zoals verstedelijking en de bouw van dammen, kunnen de kans van een overstroming verergeren door o.a. een toename van het volume afstromend regenwater als gevolg van veranderend landgebruik en bodemdaling.

Het vaststellen van het overstromingsrisico is een kritisch onderdeel van het risicomanagement van overstromingen (FRM), waaronder het bepalen van de wijze van reageren op veranderingen die in de toekomst plaats kunnen vinden (als klimaatverandering, economische verandering en bevolkingsgroei). FRM in de huidige context vraagt om statistisch robuuste benaderingen om de toenemende complexiteit van onzekerheden te kunnen meenemen. Voor een kwantitatieve benadering van FRM is het daarom noodzakelijk een probabilistische aanpak te volgen. Echter, berekeningen volgens een probabilistische aanpak vragen 2D modelberekeningen die zeer tijdrovend zijn.

Meer en meer vraagt FRM om betrokkenheid van stakeholders, die niet noodzakelijkerwijs experts zijn in risicomanagement van overstromingen. Om tot effectieve en sociaal gedragen oplossingen te komen, die tot een verlaging van het overstromingsrisico leiden, is het nodig om een brede vertegenwoordiging van stakeholders in dit proces van samen leren en samen ontwerpen, mee te nemen. Dit vereist het ter beschikking hebben van overstromingsmodellen die eenvoudig van aard zijn en daardoor toegankelijk zijn voor niet-specialisten.

Om aan bovenstaande condities te kunnen voldoen is het nodig om over een overstromingsmodel te beschikken dat snelle en voldoende accurate inschattingen kan maken van het overstromingsrisico en dat bovendien toegankelijk is voor een brede groep van stakeholders. Deze studie presenteert en demonstreert een dergelijk overstromingsmodel.

Deze studie is gestart met een gedetailleerd ééndimensionaal model van de Mekong Delta, dat geïmplementeerd is in een toegankelijk, open-source modelsysteem (SWMM) van de US-EPA. Een iteratief en vereenvoudigde model aanpak is gevolgd om snelle en accurate waterniveauvoorspellingen te kunnen uitvoeren voor een locatie aan de rivier nabij Can Tho, de grootste stad in deze delta. Deze stad is als een case-study gebruikt in dit onderzoek. Het uiteindelijke model is gekalibreerd en gevalideerd met veldwaarnemingen van rivierafvoeren en waterniveaus die zijn verzameld tijdens verschillende overstromingen. Het resultaat laat zien dat het mogelijk is waterniveaus van de rivier te simuleren met voldoende nauwkeurigheid voor eenieder willekeurige locatie in het complexe delta-rivier systeem van (het lagere deel van) de Mekong Delta met een relatief eenvoudig (teruggebracht van enkele duizenden tot enkele tientalen cross-secties) en snel (een verkorting van de simulatietijd van 1,5 uur tot ongeveer 1 minuut voor een 1-jaar simulatie) numeriek model.

Het ontwikkelde vereenvoudigd 1D model is gebruikt om waterniveaus van de rivier bij Can Tho te berekenen onder verschillende klimaat- en hydrologische condities. De modeluitkomst voor een toekomstige situatie (2050), waarbij de invloed van klimaatverandering is meegenomen (RCP 4.5 en RCP 8.5) op rivierafvoer en zeespiegelstijging (getij, zeespiegel stijging en stormen), is vergeleken met een simulatie van de huidige situatie (baseline). Voor beide periodes zijn de overstromingsfrequentie analyses uitgevoerd om de waterniveaus van de rivier te bepalen voor een scala van herhalingstijden. Extreme waterniveaus van de rivier zijn afgeleid van de resultaten voor 24 uurs tijdseries rondom de piekwaarde van iedere extreme waarde. Deze tijdseries zijn onderzocht op patronen waaruit twee typische (hydrographs) patronen zijn geïdentificeerd. De respons van deze patronen op de overstromingsniveaus in Can Tho is vervolgens kwantitatief vergeleken. De hydrograph patronen die een grote invloed hebben op een overstroming zijn gebruikt om overstromingssimulaties uit te voeren met een 1D/2D gekoppeld model in Can Tho. Dankzij de mogelijkheid om representatieve overstroming-hydrograph patronen voor de overstromingssimulaties te gebruiken, is het mogelijk gebleken om het aantal tijdrovende 1D/2D simulaties terug te brengen voor het verkrijgen van probabilistisch resultaten. Vervolgens is een regressieanalyse uitgevoerd met de uitkomsten van het 1D/2D overstromingsmodel om de relatie tussen waterniveau van de rivier en de maximale overstromingsdiepte vast te stellen voor iedere gridcel van het 2D model. De maximale overstromingsdiepte van alle gridcellen zijn geaggregeerd om overstromingskaarten te maken voor verschillende herhalingstijden, voor zowel het heden als voor 2050, als wel met en zonder het effect van bodemdaling.

De resultaten laten zien dat indien de huidige snelheid van bodemdaling zich voortzet, tegen 2050 de omvang van de overstroming (onder zowel het RCP 4.5 als het RCP 8.5 klimaat scenario) bij een herhalingstijd van 0,5 en 100 jaar zal toenemen met respectievelijk een factor 15 en 8. Echter zonder bodemdaling, zal de voorspelde toename van de omvang van de overstroming bij een herhalingstijd van 0,5 en 100 jaar tegen 2050 (onder zowel het RCP 4.5 als het RCP 8.5 klimaat scenario) beperkt blijven tot respectievelijk een verdubbeling en een verdrievoudiging.

Overstromingskaarten zijn tezamen met informatie over landgebruik (voor de huidige en toekomstige situatie), waarbij de laatste vervolgens gekoppeld is aan innudatiediepte-schadefuncties, zijn gebruikt om de te verwachte jaarlijkse schade (EAD) te berekenen. De EAD voorspellingen voor 2050 laten een stijging zien van 1,7 en 1,8 maal de EAD ten opzichte van de huidige situatie als gevolg van klimaatverandering voor respectievelijk beide scenario's RCP 4.5 en RCP 8.5 en een significante stijging van respectievelijk circa 20 en 21 maal wanneer ook de effecten van bodemdaling mee worden genomen.

Het vereenvoudigde 1D model voor de gehele Mekong Delta, de overstromings- en schadekaarten en de berekende schades verkregen uit het 1D/2D model zijn gecombineerd om een interactief, web-based hulpmiddel (*Inform*) te ontwikkelen voor het maken van een snelle bepaling van het overstromingsrisico. Dit hulpmiddel heeft als doel stakeholders te betrekken bij het ontwikkelen van effectieve en sociaal-geaccepteerde maatregelen voor het reduceren van het overstromingsrisico in Can Tho. *Inform* beschikt over de gewenste opties (zoals gebruiksvriendelijkheid) die nodig zijn voor co-design hulpmiddelen en een gebruiksvriendelijke interface. De eerste, zij het beperkte tests die zijn uitgevoerd in het kader van dit onderzoek laten zien dat *Inform* beschikt over de belangrijkste voorwaarden waaraan een dergelijk co-ontwerp tool zou moeten voldoen (e.g. ingebouwde bibilitheek met invoergegevens, flexibel en eenvoudig in gebruik, gebruiksvriendelijke interface). Bovendien neemt het ongeveer 1 minuut in beslag voor *Inform* om tijdseries voor waterniveaus te berekenen die de maximale waterniveau in de rivier bij Can Tho weergeven waarmee vervolgensde overstromings- en schadekaarten, alsmede de geschatte schade als gevolg van een willekeurige overstroming kunnen worden berekend. Deze korte rekentijden dragen bij aan de gebruiksvriendelijkheid van *Inform*.

Contents

Chapter 1

Introduction

1.1 Background and motivation

Coastal cities are among the most urbanized and populated areas of the world (Small and Nicholls, 2003; Valiela et al., 2006; Ranasinghe and Jongejan, 2018). The continued human attraction to the coast, especially in the last five decades, has resulted in the rapid expansion of settlements, urbanisation, infrastructure, economic activities, and tourism. According to statistics of low elevation coastal zones (LECZs), of the world's 20 megacities, 15 are in LECZs. RMS (Risk Management Solutions) and Lloyd's (2008) predicted that around the year 2030 more than 50% of the global population is expected to live within 100 km of the coast, especially the population living in the LECZs projected to exceed one billion (by 2050) in all SSPs (Merkens et al., 2016) and likely to reach 1.4 billion by 2060 under the high-end growth assumption (Neumann et al., 2015).

Coastal cities are facing natural disasters including flooding, which is one of the most frequently occurring and damaging natural disasters in the world (Hirabayashi et al., 2013; Kundzewicz et al., 2014; Arnell and Gosling, 2016; Alfieri et al., 2017; Forzieri et al., 2017; Mora et al., 2018). About 250 million people in the world are affected by floods every year (UNISDR, 2013), and the annual average economic losses have exceeded 40 billion USD in recent years (OECD, 2016). Coastal and delta areas are among the most flood hazard-prone areas of the world (Nicholls, 2004; Nicholls et al, 2007; Wong et al., 2014; Neumann et al., 2015; Pasquier et al., 2019), and flood intensity and frequency are already increasing, especially in coastal and delta cities, due to changes in upstream river flows, downstream sea-level and local changes in rainfall and land use (Merz et al., 2010; Balica et al., 2012; Huong and Pathirana, 2013; Chen et al., 2018; Ngo et al., 2018).

Climate change is now recognized as a major global challenge in the 21st century and beyond. Future projections indicate that climate change will have implications on the trends of extreme events (e.g. extreme precipitation, hurricanes, etc.), which may lead to increases of flooding in the future (Panagoulia and Dimou, 1997; Menzel et al., 2002; Prudhomme et al., 2013; Alfieri et al., 2015). The massive global socio-economic impacts of climate change effects in coastal areas are discussed in detail by, among others, Stern (2007), Arkema et al. (2013), Kron et al. (2012), McNamara and Keeler (2013), Johnson et al. (2015), Brown et al. (2017). For example, the economic losses due to flooding alone in coastal cities are expected to be around USD 1 Trillion by 2050 (Hallegatte et al., 2013). Besides climate change, the population increases inevitably increase water demand, which is often satisfied by excessive groundwater extraction, which, more often than not, leads to land subsidence, further exacerbating the flood hazard (due to increased inundation levels). The combination of increased flood hazard and increased population/infrastructure will, in turn, lead to an increase in flood risk in these vulnerable areas (Nicholls et al., 2007; Lenderink and Van Meijgaard, 2008; Syvitski et al., 2009; Min et al., 2011; Balica et al., 2012; Rojas et al., 2013; Wong et al., 2014; Takagi et al., 2015).

Assessing flood risk is an essential part of FRM (including managing both hazard and the potential consequences) which is becoming a critical process for adapting to future changes (for example climate change, economic change and population growth). Flood risk assessment is performed to facilitate the implementation of risk-informed measures aimed at minimizing (or mitigating) flood damage (Hall et al., 2003; Meyer et al., 2009, Bureau Reclamation, 2020; Mishra and Sinha, 2020). However, FRM and planning decisions in many parts of the world have historically utilised flood hazard or damage maps associated with one or two pre-determined return period of water levels (e.g. 100-year, 200-year). While this may have been sufficient in the past, the need to move from stationary to innovative time-dependent flood hazard and risk modelling approaches that can account for the uncertainty, arising from anthropogenic and climatic induced stressors, are rapidly increasing (Mosavi et al., 2018). Therefore, contemporary FRM is required to embrace much more statistically robust approaches. In this context, a probabilistic approach is a fundamental requirement for quantitative flood risk assessments, which aid urban planners and decision-makers to develop informed risk reduction strategies that minimize the damage caused by floods. This is especially important for coastal cities which are not only facing the impact of upstream flow changes due to human interventions (land-use change, dam construction) but also the effect of climate change on sea level (tide, sea-level rise, storm surges) as well as the effect of land subsidence in deltas.

Probabilistic flood risk assessment typically requires multiple (thousands of) river and flood model simulations to derive probabilistic flood hazard distributions in the study area. This requirement constitutes a major challenge due to the associated computational and resource demand. Although the last decade has witnessed a great improvement in computational capabilities and models, traditional modelling approaches still pose significant challenges in terms of computational time required to obtain fully probabilistic flood hazard estimates (Mcmillan and Brasington. 2007; Neal et al., 2012). Therefore, developing computationally efficient modelling approaches to circumvent this particular bottleneck remains an important challenge (Berends et al., 2018a; Berends et al., 2018b; Bomers et al., 2019a; Bomers et al., 2019b). Besides, there are some other reasons, such as a lack of hydrological data, economic data, the uncertainty of data, etc. that also make quantitative flood risk assessments challenging. Furthermore, most mitigation measures are subjected to cost/benefit analyses (CBA) before their adoption, while these CBA analyses can be expensive and time-consuming (Moore, 1995). This is also likely due to the computational cost associated with such risk-reduction quantifications, which will require re-running several fully probabilistic simulations of a 1D/2D flood modelling system in which each of the potential mitigation measures would have to be implemented.

Besides assessing flood risk, developing flood risk reduction strategies to minimise damage caused by floods are essential in FRM. A wide range of interventions is available to reduce flood risk. These interventions are often categorized according to the three layers of flood risk reduction: protection, prevention, and preparedness (Lendering et al., 2018). *In optima forma* flood risk reduction strategies consider the risk reduction potential of all interventions in the system. For the selection and implementation of flood risk reduction measures, the importance of stakeholder engagement (e.g. citizens and interest groups, businesses, officials, and decision-makers, etc.) is widely acknowledged. Stakeholders are also increasingly involved in the design phase of measures supported by userfriendly flood risk models, such as in the Dutch Room for the River program (Rijke, 2014), and SimDelta (Rijcken et al., 2012; Rijcken, 2017). The effectiveness and success of co-designing meetings have been demonstrated by the Blokkendoos tool (WL Delft hydraulics, 2003; Zhou et al., 2009) used within the Room for the River program in the Netherlands. In these meetings, stakeholders participated in the design process using 'what-if scenarios' to explore the impact of various interventions on the flood level as well as the flood risk in the study area. Effective participation of stakeholders requires a fast and accurate modelling system with a user-friendly interface, which can simulate the user's interventions and provides the output in a short time. Advanced versions of such modelling systems are also known as serious games, which are used to engage stakeholders in dialogue and activity in order to contribute their knowledge, understanding through interaction in the collaborative process in co-learning sessions (Rodela et al., 2019; Den Haan et al., 2020).

To meet the challenges discussed above, the common pre-requisite is a flood modelling system that can provide rapid and sufficiently accurate estimates of coastal city flooding and associated risks. Such a model system, especially one that facilitates thousands of simulations required for risk (and risk-reduction) quantification on a standard PC does currently not exist. This study focuses on addressing this need and uses for one case study site, Can Tho City, the largest city in the Mekong Delta, to demonstrate the value added of an innovative modelling approach which meets the requirements mentioned above.

1.2 Study area

1.2.1 The Mekong Delta

The Mekong River originates from Tibet, and flows through China, Laos, Myanmar, Thailand, Cambodia and into the East Sea of Vietnam (Fig 1.1. a). With a length of

4800 km and a mean annual flow of 475 km^3, the Mekong River ranks twelfth and tenth in the world in term of length and flow, respectively (MRC, 2005; MRC, 2010a).

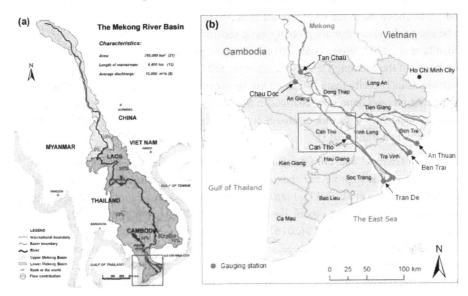

Figure 1.1. (a) Location of the Mekong Delta (Source: Mekong River Commission, Phnom Penh, Cambodia); (b) Detailed descriptions of the Mekong Delta include Can Tho city and 12 provinces as well as the location of hydrological gauging stations (revised from Kuenzer et al., (2013)). The red box in Fig. 1b shows the location of Can Tho corresponding to Fig 1.2

The Mekong Delta is the largest delta of Vietnam, and is located in the Lower Mekong River Basin (Fig 1.1.a), spanning latitudes 8°33' N and 11°01' N and longitudes 104°26' E and 106°48' E. It includes Can Tho city and the 12 provinces: Long An, Dong Thap, Vinh Long, Tien Giang, Tra Vinh, Ben Tre, Hau Giang, Soc Trang, Bac Lieu, Kien Giang, An Giang, and Ca Mau, with a total land area of about 4 million ha. The Mekong Delta has a population of approximately 17.5 million, accounting for 19 % of the country's population, while this region accounts for only 13 % of the country's area. The livelihoods of a majority of the population (85%) in the region depend on agricultural activities (Nguyen, 2008). The Mekong Delta is known as the granary of the nation and is also a key area for the production of fishery, fruit and agricultural products. Annually, it contributes about 90 % of rice, 70 % of fruit, and 60 % of fishery products in the national export turnover for each class. The economic growth of the region reached 7.39 % in 2017, up 0.49 % compared to 2016 (6.9 %). The per capita income in the Mekong Delta is about 40.2 million VND (around 1770 USD). Furthermore, the Mekong Delta is also known as one of the most biologically diverse in the world with abundant fauna and flora (e.g., fish, lizards, mammals, etc.), including rare species such as Laotian rock rat, is thought to be extinct.

The Mekong Delta has a low and flat terrain, with an average elevation of between 0.7 and 1.2 m above mean sea level (Balica et al., 2014). The Mekong Delta is also thought to be one of the areas that are globally most sensitive to the impacts of climate change (Nicolls et al., 2007; Wong et al., 2014). The most prominent reason for the climate sensitivity of the Mekong Delta is the strongly felt influence of the sea-level rise. For example, Can Tho city that is some 80 km upstream from the ocean, is still impacted by tidal variation (and hence any changes in the sea level in the future). For example, the cities drainage system cannot discharge water to the river during the high tide periods (Huong and Pathirana, 2013).

Apart from sea level rise, climate change is likely to also affect the river flow in the Mekong Delta, due to projected changes in temperature and rainfall (Hapuarachchi et al., 2008; Kingston et al., 2011). Future projections of the upstream flow in the Mekong Delta presented by, among others, Eastham et al., (2008), Hoanh et al., (2010), Västilä et al., (2010), and Hoang et al., (2016) indicate increases in river flow for all IPCC RCPs. In addition to climate change, river flow is also affected by human activities such as land-use change, upstream dam construction. The foreshadowed construction of new upstream hydropower dams is expected to vary the river flow regime (Hoanh et al., 2010; Räsänen et al., 2012), with the flow varying from year to year depending on hydropower operations (Räsänen et al., 2017). Furthermore, the Mekong Delta is also facing land subsidence (Erban et al., 2014), which results in increased flood risk in coastal and delta areas. Land subsidence is likely to increase in the future due to increased groundwater demand (Erban et al., 2014; Minderhoud et al., 2015).

While flooding (which usually occurs between July and November) in the Mekong Delta has many negative impacts (e.g., damage to property, infrastructure, and crops, loss of life) on inhabitants and economic development of the area it also has some positive impacts such as washing away salt and alum from the soil, providing supplementary fertiliser for rice fields, and increasing fish resources (Nguyen, 2008; Tran et al., 2008). The threat of flooding in future will be undoubtedly be affected by both climate change and human interventions as described above (Wassmann et al., 2004; Balica et al., 2012; Hoang et al., 2018).

This study focuses on Can Tho City, which is the cultural and economic centre of the Mekong Delta.

1.2.2 Can Tho City

Can Tho is the largest city within the Mekong Delta, with a population of around 1.6 million as of 2016, is the city located on the south bank of the Hau River, one of two major branches of the Mekong River in Vietnam, at a distance of about 80 km upstream of the sea (Fig 1.2). Can Tho has a dense system of rivers and canals, and therefore, it is also known as "the municipality of river water region".

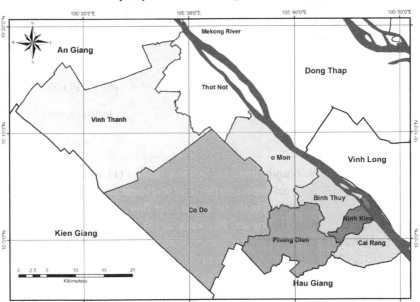

Figure 1.2. A map of Can Tho city (in the red box of Fig. 1b). The area shown here extends much beyond the urban centre. The urban area is largely in the lower right corner of the diagram

Can Tho is a dynamic city that is emerging as an economic centre and is expected to play an important role in the Mekong Delta and the adjacent international regions in future (Huong and Pathirana, 2013; MDP, 2013). With its strategic location, Can Tho city is expected to witness exponential growth in the next several decades (Huong and Pathirana, 2013; NIURP, 2010). The rapid urbanization and growth of population in Can Tho will likely lead to a significant change in land use within and the surrounding area of the city (Huong and Pathirana, 2013). This may result in substantial changes in the urban water cycle, increasing flood frequency and water pollution, all of which will further increase the city's already high risk of flooding in the coming decades.

Flooding is frequent in Can Tho City, which leads to significant impacts on environment, economy and society. Flooding occurs at least 2-3 times a year during the monsoon season, with floodwater depths ranging from a few centimetres up to 20-30 cm (even up to 70-80 cm in some places such as Ninh Kieu district). According to the Southern Hydrometeorology Station, in 2011, the maximum water level reached 2.15 m relative to mean sea level. Table 1.1 shows the water level in Can Tho city in flooding years.

Table 1.1. Water level and discharge in flooding years in Can Tho (relative to MSL) since 2000 (Source: Adapted from (CCCO & ISET 2015))

Year	Date	Highest Water Level (cm)	Average Discharge in Can Tho (m³/s)	
			On Day of Flooding	Highest Flow in Month of Flooding
2000	30/09	179	13,000	(23/09) 17,700
2011	27/10	215	16,100	(05/10) 19,600
2013	20/10	213	12,290	(30/10) 18,180
2014	10/10	208	-	-

Flooding in the city is significantly impacted by three factors: (1) sea level (tide, storm surge, sea-level rise) has a direct impact on the river level near Can Tho and hence the dynamics of flooding; (2) Changes in the upstream river flow, which is impacted by the change in the rainfall regime (Huong and Pathirana, 2013), land-use change, construction of hydraulic structures such as dams (Lauri et al., 2012; Piman et al., 2013; Räsänen et al., 2017; Pokhrel et al., 2018); and (3) Urban hydrology (changes in local rainfall, land use, etc.). To represent the first two factors above, a physically-based simulation model of the river system of the Mekong Delta is essential. With such a model it is possible to use sea-level changes and upstream flow changes as boundary conditions to ascertain their compound impact on the river water level near Can Tho city and therefore the impact on floods in the city.

To cope and adapt to the flooding situation in Can Tho, Can Tho's municipal authorities have reconstructed a dyke system to protect the urban area from river and tidal flooding. In addition, some pumping stations have been built at several locations in the city and combined with the existing urban drainage system to drain water. However, the urban drainage system is not functioning optimally (Huong and Pathirana, 2012). Individual households use ad-hoc adaptations measures such as raising floor levels, building small temporary dikes in front of their houses during the floods to enhance their coping capacity (Birkmann et al., 2010; DWF 2011; Radhakrishnan et al., 2018a).

1.3 Hydraulic models and applications

1.3.1 Hydraulic models for rivers

With the rapid development of computing facilities over the last decades, hydraulic models have been developed to address the complex real-world hydraulic problems. Many hydraulic models for different purposes like floodplain modelling, river modelling and drainage modelling, were developed such as MIKE, ISIS (Flood

Modeller), Hydro-GIS, Delft3D, HEC-RAS, SOBEK, SWMM, PCSWMM, LISFLOOD-FP, etc. (Domeneghetti et al., 2012; Tri et al., 2012).

Hydraulic models are categorised according to their dimensionality. They represent the spatial domain and flow processes, one-, two- or even three-dimensional. The different types of models are selected depending on the issue being investigated (Hunter et al., 2006).

In this study, one-dimensional (1D) and two-dimensional (2D) hydraulic models are used to derive estimates of water levels and inundation levels (2D) in the case study area (Can Tho).

The flow in natural rivers and floodplains comprises 1D and 2D unsteady flow, which is represented by the Saint-Venant equations (also referred to as the shallow water equations). The Saint-Venant equations are derived from the conversion of mass and momentum (Toombes and Chanson, 2011), expressed as equations 1.1 and 1.2.

$$\int_S \dot{m} dA = \frac{dM}{dt} \tag{1.1}$$

$$\frac{d}{dt}\left(m \times \vec{V}\right) = \sum \vec{F} \tag{1.2}$$

where:

$\dot{m}$: the mass flow rate across the control surface (kg/s)

A: the flow area (m^2)

M: the mass within the control volume (kg)

t: time (s)

$\vec{F}$: the force (vector) acting on the control volume

$\vec{V}$: the velocity (vector) of the control volume

For a Newtonian fluid (the viscous stress is proportional to the local strain rate due to fluid movement) density (ρ) and viscosity (μ) of the fluid are constant, equations (1.1) and (1.2) in a cartesian coordinate system, can be written as below:

$$\frac{\partial u}{\partial x} + \frac{\partial v}{\partial y} + \frac{\partial w}{\partial z} = 0 \tag{1.3}$$

$$\rho\left(\frac{\partial u}{\partial t} + u\frac{\partial u}{\partial x} + v\frac{\partial u}{\partial y} + w\frac{\partial u}{\partial z}\right) = -\frac{\partial p}{\partial x} + \mu\left(\frac{\partial^2 u}{\partial x^2} + \frac{\partial^2 u}{\partial y^2} + \frac{\partial^2 u}{\partial z^2}\right) + \rho g_x \tag{1.4}$$

$$\rho\left(\frac{\partial v}{\partial t}+u\frac{\partial v}{\partial x}+v\frac{\partial v}{\partial y}+w\frac{\partial v}{\partial z}\right)=-\frac{\partial p}{\partial y}+\mu\left(\frac{\partial^2 v}{\partial x^2}+\frac{\partial^2 v}{\partial y^2}+\frac{\partial^2 v}{\partial z^2}\right)+\rho g_y$$

(1.5)

$$\rho\left(\frac{\partial w}{\partial t}+u\frac{\partial w}{\partial x}+v\frac{\partial w}{\partial y}+w\frac{\partial w}{\partial z}\right)=-\frac{\partial p}{\partial z}+\mu\left(\frac{\partial^2 w}{\partial x^2}+\frac{\partial^2 w}{\partial y^2}+\frac{\partial^2 w}{\partial z^2}\right)+\rho g_w$$

(1.6)

Where:

u, v, w: flow velocity in the x, y, z directions (m/s)

p: the pressure (N/m^2)

g_x, g_y, g_z: gravitational acceleration in the x, y, z directions (m/s^2)

ρ: the density of the fluid (kg/m^3)

μ: the viscosity of the fluid (m/s^2)

Equations 1.4, 1.5 and 1.6 are known as the Navier-Stokes equations.

1D hydraulic models

1D hydraulic models can be classified as steady or unsteady. Steady-state numerical models are based on the energy equation, while unsteady models are based on the Saint-Venant equations (Toombes and Chanson, 2011).

The assumptions used in the 1D solution of the Saint-Venant equations include:

- The fluid is incompressible.

- The flow is 1D.

- The streamline curvature is small.

- The hydrostatic pressure prevails and vertical accelerations are negligible.

- The channel bottom slope is small.

The Saint-Venant equations can be presented as equation 1.7 and 1.8.

$$\frac{\partial A}{\partial t}+\frac{\partial(vA)}{\partial x}=0$$

(1.7)

$$\frac{\partial Q}{\partial t}+\frac{\partial}{\partial x}(uQ)+gA\left(\frac{\partial h}{\partial x}-S_0\right)+gAS_f=0$$

(1.8)

Where:

A: wetted cross-section area (m^2)

Q: flow discharge (m^3/s)

u: longitudinal flow velocity (m/s)

g: gravitational acceleration (m/s^2)

S_0: bed slope

S_f: friction slope

1D hydraulic models need initial boundary conditions at the entire model domain and boundary conditions, at upstream (e.g. flow hydrograph, stage hydrograph) and downstream (e.g. observed water level surface, normal depth, flow/stage hydrograph, rating-curve) for the entire modelling period. 1D hydraulic models are used to provide estimates of water depth hydrograph and velocity at locations in the system.

A number of 1D hydraulic models are widely used for rivers such as MIKE11, HEC-RAS, ISIS (Flood modeller), SWMM, InfoWorks, SOBEK, etc.

2D numerical hydraulic models

2D hydraulic models have been used for modelling of floodplains, urban, coastal and marine situations, etc. where have complex flow patterns. 2D hydraulic models use numerical solutions based on the Saint-Venant equations (as referred to the depth-averaged shallow water equations).

When the flow can not be simplified as one-dimensional (e.g. wide river cross sections, flood plains), the 2D form of the 2D Saint-Venant equations (referred to as the shallow water equations) can be used under similar assumptions as those of 1D equations above. These are expressed in equations 1.9, 1.10, and 1.11.

$$\frac{\partial h}{\partial t} + \frac{\partial (hu)}{\partial x} + \frac{\partial (hv)}{\partial y} = 0 \tag{1.9}$$

$$\frac{\partial (hu)}{\partial t} + \frac{\partial}{\partial x}\left(hu^2 + \frac{1}{2}gh^2\right) + \frac{\partial (huv)}{\partial y} = gh(S_{0x} - S_{fx}) \tag{1.10}$$

$$\frac{\partial (hv)}{\partial t} + \frac{\partial}{\partial y}\left(hv^2 + \frac{1}{2}gh^2\right) + \frac{\partial (huv)}{\partial x} = gh(S_{0y} - S_{fy}) \tag{1.11}$$

Where:

u: velocity in the x direction (m/s)

v: velocity in the y direction (m/s)

g: gravitational acceleration (m/s^2)

h: water depth (m)

S_{0x}, S_{0y}: bed slopes in the x and y directions

S_{fx}, S_{fy}: friction slope in the x and y directions

In addition to the initial conditions over the model domain, 2D models need specified boundary conditions over the entire model boundaries, which may include the sides of the domain as well.

TUFLOW, MIKE21, SOBEK2D, InfoWorks-2D, Delft 3D, VRSAP, and HydroGIS, PC-SWMM, etc. are commonly used 2D hydraulic models.

Coupling of 1D/2D numerical models

In recent years, the coupling between 1D and 2D hydraulic models has become popular (Seyoum et al., 2012; Adeogun et al., 2015; Leandro et al., 2016; Fan et al., 2017; Bomers et al., 2019c; Pasquier et al., 2019; Dasallas et al., 2020; Cardoso et al., 2020). This helps modellers to take advantages of the strengths of each model. 1D hydraulic models can simulate complex structures such as drainage networks, sewers, spills, reservoirs, bridges, etc. 2D hydraulic models can resolve complex floodplain flow and generate inundation levels. The combination of 1D and 2D hydraulic models not only enhances computational efficiency but also improves the accuracy of model results (Leandro et al., 2016; Fan et al., 2017; Dasallas et al., 2020).

Several modelling systems such as HEC-RAS, SOBEK 1D 2D, LISFLOOD-FP, MIKE FLOOD, PCSWMM, etc. can handle coupled 1D-2D hydraulic modelling. For example, MIKE FLOOD has been developed to conform to several types of links (e.g. standard link, lateral link and structure link) between 1D MIKE 11 and 2D MIKE 21. In HEC-RAS, lateral structures are used to connect the 1D river reach to the 2D area, while bottom orifices are used to connect junctions in the 1D model with the 2D mesh in PCSWMM. These links allow exchanging flow between the 1D and 2D models.

1.3.2 Previous applications of hydraulic river models for the Mekong Delta

Over the last decades, several modelling studies were conducted for the Mekong Delta in Vietnam with the different purposes.

Wassmann et al. (2004) used the "Vietnam River System and Plains" (VRSAP) model to assess the water level changes in the Mekong Delta due to the impacts of sea-level rise. The VRSAP model was developed by the Sub-Institute for Water Resources Planning, Ministry of Agriculture and Rural Development, Vietnam (Khue, 1986). The VRSAP model for the whole Mekong Delta included 1505 nodes, 2111 segments, and 555 storage plains. It was calibrated with hydrological data obtained in 1996, and then used to predict water levels for years 2030 and 2070. This model projected water levels for the Mekong Delta when flooding is presently high (from August to November) for sea-level rise projections of 20 cm (in 2030) and 45 cm (in 2070), respectively. The results of this study indicated that the average increase in August water levels corresponding to the two considered scenario were 14.1 cm and 32.2 cm, respectively, relative to 1996. In October, the increase in water levels were found to be weaker due to high discharge from upstream, but the average increases for the two considered scenarios were still high at 11.9 cm and 27.4 cm, respectively.

Le et al., (2007) used the HydroGis model, developed by the Ministry of Natural Resource and Environment of Vietnam (MONRE), which combines a hydrodynamic model and GIS tools (Le et al., 2005) to evaluate changes in flooding in the Mekong Delta due to the combined effect of upstream river flows, storm surge, sea-level rise, estuarine siltation, and hydraulic structures. The HydroGis model for the entire Mekong Delta comprised 13,262 cross sections, 2535 flood cells, and 467 hydraulic structure like sewers, sluices, and bridges. This study indicated that the flood levels depend on the combined impacts of Mekong River flows, storm surges and sea-level rise as well as the construction of upstream dams. Water levels in the delta were predicted to increase from 5 cm to 200 cm corresponding each scenario. Additionally, this study predicted that the construction of upstream dams would cause siltation the Hau river estuary and followed by the increase by up 2 m in peak flood levels in the Mekong Delta.

HR Wallingford and Halcrow (UK) developed a 1D ISIS model for the Mekong Delta to assist the Mekong River Commission (MRC) in managing and using water resources in the Mekong River Basin (Van et al., 2012). Currently, this model is maintained by the MRC. MRC's detailed modelling studies for Cambodia and the Vietnamese Mekong Delta have used the results of this model as boundary conditions. Van et al., (2012) used this model to study changes of flood characteristics under the impacts of upstream development and sea-level rise. The 1-D ISIS model for the Mekong Delta comprised 3036 cross-sections representing 8619 km system of river and canals, 538 junctions, 749 floodplain units, 193 spillways, 409 reservoirs, and 29 sluices. This study used the flood event data of the year 2000 to validate the model, and subsequently used it to predict potential changes in flooding for the year 2050. The results showed that the flood hazard in 2050 may become more severe along the coastal area. Additionally, future (2050) inundation levels in the upstream Mekong Delta were projected to be lower and shorter than in 2000, while along the coastal areas would be higher and longer.

All the models applied in the aforementioned studies are models with a high level of detail (many cross-sections and hydraulic structures (e.g. sewers, sluices, reservoirs, etc.)). If these models are well-calibrated, they can provide accurate results. However, these models do take a long time for extended period simulations, owing to the large and complex application domain. For example, the time for the aforementioned ISIS model of Mekong Delta to complete a one-year simulation is about 90 min (on a single core on an Intel(R) Core(TM) i5-4210M CPU @ 2.6 GHz processor in a computer with 8.0 GB of memory (RAM)). With such a computational time, it is impractical to execute the thousands of individual simulations required for probabilistic flood forecasting with this model as required for risk assessments and subsequent risk-informed decision making.

1.4 Main drivers that affect the flooding situation in the Mekong Delta and Can Tho

According to the above description of the case study area, three drivers were identified as the main contributors to increase the flood level in the Mekong Delta, including climate change; human interventions (land-use change, dam construction) and land subsidence. The following sub-sections summarise the available future projections on climate change driven variations in rainfall, sea-level rise, storm surge and river flow; the changes in river flow due to dam construction; and land subsidence rate.

1.4.1 Future projections of climate change driven variations in main drivers of flooding in the Mekong Delta.

In previous studies, climate change observations in the Lower Mekong Basin (including the Mekong Delta) indicate changes in rainfall distribution, sea-level rise and river flow (Eastham et al., 2008; Hoanh et al., 2010; Västilä et al., 2010; Hoang et al., 2016; MONRE 2016).

In 2014, the Intergovernmental Panel on Climate Change (IPCC) released the 5th assessment report (AR5) on climate change, which included projections about climate change and sea-level rise scenarios for all regions in the world including Southeast Asia region.

Based on the results of IPCC for Southeast Asia, MONRE used a combination of hydrological, sea level, and topographic data; coupled atmosphere-ocean and global and regional climate models; and the studies of the Viet Nam Panel on Climate Change and the Institute of Meteorology, Hydrology and Climate Change to obtain more region specific projections for climate change driven variations in rainfall, sea-level and storm surge for all regions in Vietnam, including the Mekong Delta.

Rainfall

MONRE (2016) provides the following projections for rainfall in Vietnam by late 21st century: annual rainfall would increase by 5-15 % over most parts in Vietnam for the RCP 4.5 scenario and by over 20 % for RCP 8.5 scenario over most parts of the country except a part of the South. All regions in Vietnam will experience an increase by 10-70

% in average maximum daily rainfall relative to the reference period (1996-2005). Table 1.2 presents projected changes in annual rainfall of the 13 provinces in the Mekong Delta for scenarios RCP 4.5 and RCP 8.5 corresponding to 3 specific periods compared to the baseline period of 1986-2005.

Table 1.2. Projected changes in annual rainfall (%) of the 13 provinces in the Mekong Delta relative to the 1996-2005 baseline (values in parentheses are the range of rainfall changes from climate model results with the lower boundary of 20 % and the upper boundary of 80 %, respectively) (Source: MONRE (2016))

Province, City	RCP4.5 scenarios			RCP8.5 scenarios		
	2016-2035	2046-2065	2080-2099	2016-2035	2046-2065	2080-2099
Long An	11.7	20.6	16.7	12.8	16.1	19.9
	(4.1÷18.4)	(7.7÷33.9)	(2.8÷29.1)	(5.8÷19.2)	(9.1÷23.5)	(11.5÷28.3)
Dong Thap	10.0	17.9	17.2	11.0	16.2	23.7
	(4.7÷15.2)	(8.8÷28.1)	(5.2÷28.5)	(4.3÷17.5)	(10.6÷22.3)	(15.5÷32.1)
Vinh Long	6.2	9.1	11.1	7.6	11.8	13.4
	(2.4÷10.2)	(1.3÷17.8)	(0.5÷21.9)	(2.6÷13.3)	(7.1÷17.1)	(4.4÷23.8)
Tien Giang	13.7	17.1	16.1	12.7	18.0	20.9
	(8.5÷18.8)	(7.2÷28.4)	(2.6÷28.9)	(6.4÷18.8)	(10.5÷25.9)	(10.4÷32.4)
Tra Vinh	10.9	15.7	17.7	11.4	14.6	18.2
	(4.8÷16.4)	(5.6÷26.9)	(4.0÷30.1)	(5.5÷17.6)	(8.3÷21.6)	(9.1÷28.1)
Ben Tre	17.0	18.2	21.2	14.7	18.1	21.8
	(10.0÷23.3)	(7.5÷30.5)	(7.6÷33.7)	(9.6÷19.9)	(11.2÷25.7)	(11.2÷33.1)
Hau Giang	4.9	4.5	7.4	3.8	8.6	9.8
	(2.0÷7.9)	(-2.4÷11.8)	(-1.7÷17.1)	(0.1÷7.9)	(4.3÷13.5)	(0.4÷21.1)
Can Tho	10.5	13.7	15.1	10.7	18.3	21.2
	(6.5÷14.5)	(4.4÷23.7)	(2.7÷26.7)	(4.1÷18.1)	(13.4÷23.5)	(12.2÷30.8)
Soc Trang	11.1	10.6	14.0	10.6	15.4	18.4
	(7.1÷15.1)	(2.1÷19.6)	(4.1÷23.6)	(5.2÷16.6)	(10.3÷20.7)	(9.7÷28.4)
Bac Lieu	9.6	11.0	13.6	11.8	16.5	18.0
	(5.1÷13.8)	(2.2÷20.6)	(4.2÷22.9)	(6.3÷18.1)	(10.0÷23.4)	(8.4÷29.1)

Province, City	RCP4.5 scenarios			RCP8.5 scenarios		
	2016-2035	2046-2065	2080-2099	2016-2035	2046-2065	2080-2099
Kien Giang	4.9 (0.1÷10.2)	9.2 (0.7÷18.5)	17.0 (2.2÷31.9)	6.5 (-1.3÷14.7)	14.4 (7.2÷21.8)	15.4 (4.3÷28.1)
An Giang	4.7 (-0.4÷9.5)	13.1 (3.7÷23.4)	14.1 (0.4÷26.5)	8.2 (1.4÷15.2)	11.1 (5.3÷17.4)	14.7 (6.6÷23.5)
Ca Mau	8.4 (2.0÷14.1)	5.8 (-2.5÷14.8)	9.6 (-0.4÷19.6)	6.7 (2.1÷11.8)	10.8 (6.1÷16.1)	12.6 (3.8÷22.8)

Sea level rise

According to MONRE (2016), projections for sea level rise at the Mekong River mouths (Mui Ke Ga-Mui Ca Mau) by late 21st century relative to the period 1986-2005 for scenarios RCP 4.5 and RCP 8.5 are as presented in Table 1.3.

Table 1.3. Sea-level rise scenarios for the Mekong Delta (Mui Ke Ga-Mui Ca Mau) (cm) (values in parentheses are levels of confidence of 5 % and 95 %, respectively) (Source: MONRE (2016))

Timeline	RCP Scenario	
	RCP4.5	RCP8.5
2030	12 (7÷18)	12 (8÷17)
2040	17 (10÷25)	18 (12÷26)
2050	22 (13÷32)	25 (16÷35)
2060	28 (17÷40)	32 (21÷46)
2070	33 (20÷49)	41 (27÷59)
2080	40 (24÷58)	51 (33÷73)
2090	46 (28÷67)	61 (41÷88)
2100	53 (32÷77)	73 (48÷105)

Storm surge

Storm surge is an anomalous rise in sea level due to the high winds and reduced atmospheric pressure associated with typhoons. Storm surge occurs not much but can cause severe flooding in the coastal areas due to sudden high water level rise.

The highest storm surge that has been recorded in the estuaries of the Mekong Delta is over 200 cm, and is predicted to be more than 270 cm in the future (MONRE, 2016). At present, there is no detailed process based model application for the prediction of future storm surge along the Vietnam coast. However, based on a global application of the Delft3D-FM unstructured grid model, Muis et al. (2016) provided a 36 year hindcast of storm surge all around the world, which could be used to obtain a robust cumulative distribution function of storm surge in the vicinity of the Mekong Delta.

Mekong River discharge

Table 1.4 summarises the results of previous studies on the projected change in Mekong river discharge at Kratie under different climate scenarios.

Table 1.4. Projections of annual Mekong river discharge change (%) at Kratie under different climate scenarios

Author	Climate scenario				Predicted time	Baseline period
	A2	B2	RCP4.5	RCP8.5		
Eastham et al. (2008)	22				2030	1951÷2000
Hoanh et al. (2010)	10÷13	5÷9			2010÷2050	2010÷2050
Västilä et al. (2010)	4				2010÷2049	1995÷2004
Hoang et al. (2016)			3÷8	-7÷11	2050	1971÷2000

1.4.2 Projected change in river flow due to dam construction in the upstream part of the Mekong Delta.

The construction of upstream hydropower dams is expected to vary the river flow regime of the Mekong Delta. At Chiang Saen (Thailand), the discharge is predicted to rise by 60-90 % in the dry season and to fall by 17-22 % in the wet season (Hoanh et al., 2010; Räsänen et al., 2012). However, the hydropower operations have considerably modified the river discharges since 2011 and the largest changes were observed in 2014. At Chiang Saen, the observed discharge increased by 121-187 % in March-May 2014 and decreased by 32-46 % in July-August 2014 compared to average discharges. The respective changes at Kratie (Cambodia) were 41-74 % increase in March-May 2014 and 0-6 % decrease in July-August 2014 (Räsänen et al., 2017). Hoang et al. 2019

17

predicted that river flows would increase 63 % in the dry season and decrease 15 % in the wet season at Kratie for the 2050s period for hydropower development scenario includes 126 dams on the mainstream and tributaries of the Mekong.

1.4.3 Land subsidence in the Mekong Delta

Erban at al. (2014) used observed groundwater level data during the period 1995-2010 from monitoring wells to show that the hydraulic head declined by about 0.3 m/yr on average as a result of excessive groundwater extraction. This process leads to a compaction of the sedimentary layers and causes land subsidence at an average rate of 1.6 cm/yr. This rate is consistent with the measured results of InSAR (interferometric synthetic aperture radar). They also predicted land subsidence of about 0.88 m (0.35-1.4 m) by 2050 if the groundwater exploitation continues at above rate.

In more recent studies by Minderhoud et al, the average subsidence rate in the Mekong Delta was estimated to be 1.1 cm/yr (Minderhoud et al., 2017), and increasing to 1.31 cm/yr corresponding to B2 scenario (no-mitigation with a steady annual increase of 4% of the 2018 volume) (Minderhoud et al., 2020). However, the projected rate is likely to increase in the future due to increased groundwater demand (Minderhoud et al., 2017; Minderhoud et al., 2020).

1.5 Previous flood risk assessments studies.

Over the last decades, a few studies have been carried out to assess the flood hazard and flood risk in Can Tho.

Apel et al. (2016) developed probabilistic flood maps for fluvial hazard, pluvial hazard and combined fluvial and pluvial hazard for Can Tho city for the different quantiles (5%, 50% and 95%) of flood events corresponding to selected annual probabilities of non-exceedance (p) (e.g. 0.5, 0.8, 0.9, 0.95, 0.98, 0.99). Probabilistic fluvial flood hazard maps for Can Tho were developed based on the flood probabilities at Kratie, which were determined based on a flood frequency analysis using annual extreme discharge and flood volume at Kratie instead of Can Tho. In contrast, probabilistic pluvial flood hazard maps were developed based on the probability of local rainfall at Can Tho.

Their study indicated that with additional rainfall in a short period of time, the fluvial processes are not significantly altered. Their computed inundated area corresponding to 50 % quantile map for $p = 0.99$ for fluvial, pluvial, combined fluvial and pluvial is 5.29, 6.43 and 7.14 km^2, respectively.

Chinh et al. (2017) used probabilistic flood hazard maps produced by Apel et al. (2016) combined with multi-variable (water depth, flood duration and floor space of building) flood loss models for residential buildings and contents to assess flood risk for private households in Can Tho. The computed EAD to residential buildings and contents for the urban centre area of Can Tho City corresponding to 50 % quantile hazard maps for fluvial, pluvial, combined fluvial and pluvial were about 1.90, 1.49 and 3.34 million USD.

1.6 Socially and economically acceptable flood risk reduction measures

Flood risk management and its various interventions can not eliminate risk but reduce it. Particularly with the case of managing non-catastrophic floods, like those of Can-Tho, it is necessary to look at the cost involved in the flood risk reduction measures against the expected reduction of risk. However, this is not the only consideration in flood risk management. The capacity of a society and an economy to address flood risk is also important. The literature on flood resilience identifies three types of capacities, namely, "capacity to resist," "capacity to absorb and recover" and "capacity to transform and adapt" (Hegger et al., 2016). The level of protection provided by flood hazard reduction interventions increases the capacity to resist by reducing the hazard. Economic, ecological and social factors also influence the capacities of "absorb and recover" and "transform and adapt". For example aspects like the presence of flood awareness, insurance, early warning and crisis management contribute to the capacity to absorb and recover. An example for transform and adapt the type of resilience is the capability of institutions and communities to adopt new approaches and perspectives. What is socially and economically acceptable flood risk would therefore depend on these factors.

Can Tho city encounters many small flooding events every year. This has influenced the behaviour of the inhabitants and has resulted in different adaptation measures they have introduced in an organic and ad-hoc fashion. This is an example of a "absorb and recover" strategy. Radhakrishnan et al. (2018a), SCE (2013) and Birkman (2010), among others, list various coping actions like elevating the flood of flood-prone houses, building low flood barriers at the house thresholds, etc. However comprehensive flood risk management actions would, over the long-term reduce these actions. For example, if a neighbourhood is protected from frequent floods, it's likely that new constructions in that neighbourhood would not implement the coping mechanisms and existing coping mechanisms would not be maintained. However, this does not mean that the flood risk is not there: High return period, extreme events would still occur causing economic damage. Therefore, in contrast to Radhakrishnan et al. (2018a), this study does not consider the coping capacity of the inhabitants when calculating flood risk.

This study aims to provide estimates related to the flood risk increase due to Climate Change-driven forcing. It provides indications of the flood risk increase due to different climate scenarios. Any flood risk management strategy that will be implemented has to pay attention to these changes as these will change what is socially and economically acceptable in the future. In addition to the economic loss and nuisance, climate change may even transform floods in Can Tho into catastrophic events in the future.

1.7 Interactive tools used in FRM

In recent years, the participation of stakeholders (citizens and interest groups, businesses, officials, and decision-makers, etc.) has become more popular for co-designing FRM strategies, usually done via one or more dedicated co-designing meetings (Thaler and Levin-Keitel, 2016; Edelenbos et al., 2017; Leskens et al., 2017, Den Haan et al, 2020). In these meetings, computer models are commonly used to aid the planning process, especially interactive tools with a simplified, user-friendly interface with which stakeholders can explore several 'what-if' scenarios and be actively involved in identifying specific adaptation measures to be considered. The 'Blokkendoos' (Dutch for 'box of building blocks') is one such tool, which has been developed by WL Delft (now Deltares). The tool can show the effects of Room for the River measures on the high water level, agriculture, nature, people and the cost of these measures. In this way, the tool helps policy makers to make socially and economically acceptable decisions.

In addition to the Blokkendoos, currently, there are several online tools (web-based platform) to estimate the current and future fluvial and coastal flood risk for every country, state and river basin over the world (Aqueduct Floods tool developed by Deltares and partners) (Deltares, 2020); and to provide information on the long term flood risk across locations in The Netherlands and the UK including possible causes of flooding and how to manage this risk (Rijkswaterstaat, 2020; UK Environment Agency, 2020). However, these tools do not consider local FRM interventions at the level of a city. This requires more detailed and high resolution data. Therefore these tools do not allow explorative analyses and planning exercises with stakeholders to support strategic urban FRM, to conduct quick checks about the effectiveness of design of local FRM measures, to foster community-level participatory planning processes on FRM.

As flood risk calculations are computationally intensive and time-consuming there have been attempts to reduce the computational loading through the schematization of the urban flood plain under consideration (e.g. Kind (2014); Custer (2015); Vuik et al (2016); and Dupits (2017)). There are also recent approaches such as Berchum et.al (2020) which use a general flood risk screening modelling tool (FLORES) that uses simplified hydraulic formulas and hydrological balances, instead of detailed process-based hydraulic models such as PCSWMM or MIKE URBAN. FLORES is designed as a screening tool for the selection of FRM strategies for urban areas along the coast at a conceptual level. This broad-level screening can be based on parameters such as risk reduction, construction cost and reduction of the affected population. Further, the emphasis of FLORES is on reducing as many local inputs as possible and also to utilize open source data sets. These approximations enable the users to cut down on the data inputs and screen the flood risk reduction strategies (FRRS) across scenarios at a conceptual level.

1.8 Research objective and questions

The overarching objective of this study is to develop and demonstrate an efficient (defined here as fast, but providing sufficiently accurate results at a low computational cost) flood modelling approach to provide rapid and sufficiently accurate estimates of coastal city flooding and associated risk for present and future. In addition, the research aims at using the model outcomes to develop and pilot a co-designing interactive tool to determine acceptable flood risk reduction measures to enhance stakeholder participation in the selection process.

To achieve the above objective, this study seeks to answer the following questions by developing an efficient modelling system approach.

- To what extent is it possible to substantially simplify a complex river network model without compromising the accuracy of prediction at a specific point of interest?

- How can the computational cost of quantitative flood risk assessments which take into account the uncertainty of natural process (e.g. climate change, land subsidence) be substantially reduced?

- What are the challenges in creating interactive tools aiming to support stakeholders in co-designing flood risk reduction measures? How can an interactive tool address potential challenges that users may face during a co-design process?

1.9 Thesis outline

This dissertation is structured into six chapters, as summarized below:

Chapter 1 constitutes the introduction to this research. It consists of the research background and motivation, a description of the study area, followed by a short review on hydraulic models and applications, a summary of main drivers that affect the flooding situation in the Mekong Delta and Can Tho, then a summary of previous studies on flood risk assessment in Can Tho and a short review on interactive tools used in FRM. This chapter also introduces the overarching and specific research objectives, and questions considered in this study.

Chapter 2 presents the methodology to develop a simplified 1D hydrodynamic model for the entire Mekong Delta and its application to support probabilistic flood forecasting for Can Tho in the future, which takes into account for forcing and system uncertainty.

Chapter 3 presents an efficient combination between the simplified 1D hydrodynamic model for the entire Mekong Delta obtained in Chapter 2 with a detailed 1D/2D coupled model for the urban centre of Can Tho city to develop probabilistic flood hazard maps for the study area for present-day and future under different scenarios taking into account the impact of climate change forcing (river flow, sea-level rise, storm surge) and the effect of land subsidence. This chapter provides the method and its application to achieve the goal. This chapter also provides flood hazard maps for the urban centre of Can Tho city corresponding to a range of return periods (0.5 yr, 1 yr, 2 yr, 5 yr, 10yr, 20 yr, 50 yr and 100 yr) of water level for present-day and for 2050 under RCP 4.5 and RCP 8.5 with and without land subsidence.

Chapter 4 presents assessments of flood damage and risk for the urban centre of Can Tho city for baseline and future under different scenarios. This chapter also explores the possibility of reusing the flood hazard calculations of EAD assessment capture flood hazard profile for plausible river water levels without performing additional expensive 2D simulations and then compute flood damages in the study area caused by these plausible river water levels. The products obtained from the reusing above (e.g. flood hazard maps, flood damage maps and estimated flood damages) are used as the database for an interactive co-design tool which will be introduced in Chapter 5.

Chapter 5 presents challenges in creating co-designing interactive tools to support FRM. This chapter also presents an interactive, web-based tool – *Inform*, which was developed with the aim of supporting rapid flood risk assessment and facilitating a co-designing approach for risk reduction measures in the urban centre of Can Tho city (Ninh Kieu district) among stakeholders. Pilot testing and evaluation of *Inform* is presented, followed by a discussion on additional information that may be useful for further applications of the tool.

Chapter 6 presents the general conclusions of this research. It also outlines limitations associated with this study and provides suggestions for future research.

Chapter 2

Development of an effective modelling approach to support probabilistic flood forecasting in Can Tho city, Mekong Delta, Vietnam[1]

[1] This chapter is partially based on Ngo, H., Pathirana, A., Zevenbergen, C., Ranasinghe, R.: An Effective Modelling Approach to Support Probabilistic Flood Forecasting in Coastal Cities – Case Study: Can Tho, Mekong Delta, Vietnam, J. Mar. Sci. Eng., 1–19, doi:10.3390/jmse6020055, 2018.

2.1 Introduction

Flooding has serious negative impacts on human activities and properties in coastal cities which was amply reflected by Hallegatte et al., (2013), who predicted that the economic losses due to sea-level rise driven flooding alone in coastal cities are expected to be around USD 1 Trillion by 2050. This escalation of damage will be caused by many reasons. On the one hand, cities, particularly in the global south, are undergoing rapid land-use change due to population growth and increasing industrialization (Suriya and Mudgal, 2012; Gorgoglione et al., 2016; Zope et al., 2016). On the other, coastal and estuarine cities are threatened by increasing water levels due to both sea-level rise and changes in the upstream flow patterns (Huong and Pathirana, 2013). On top of these, there is the possibility that flooding might increase due to the local rainfall regime connected to both global climate change and the local land-use driven microclimate changes (Pathirana et al., 2014).

The ability to predict the changes in the flood hazard and risk under a variety of internal (e.g., land use) and external (e.g., climate) scenarios play an important role in FRM under a rapidly changing environment. Traditional modelling applications in the domain of FRM typically involved developing detailed flood models and conducting a handful of flood simulation exercises. Such an approach works adequately under the assumption that the forcing variables and the model conditions can be represented by a limited repertoire of scenarios – in other words ignoring the uncertainty of those parameters. Due to rapid and uncertain changes in the forcing parameters, particularly with regards to the internal ones, contemporary FRM is required to embrace much more statistically robust approaches. In this context, probabilistic forecasts are a fundamental requirement for quantitative flood risk assessments, aiding urban planners and decision-makers to develop informed risk reduction strategies that minimize the damage caused by floods. This is especially important for coastal cities which are not only facing the impact of climate change driven and human induced (e.g. upstream dam construction) variations in upstream flow changes but also sea level rise and climate change driven variations in storm surges as well as other natural processes such as land subsidence in deltas.

Hydraulic models that simulate river flow and flooding are generally used to estimate water level changes under various future scenarios. With the rapid development of computing facilities over the last few decades, many different hydrodynamic models (e.g., MIKE11, ISIS (now is Flood Modeller), HEC-RAS, SOBEK, LISFLOOD-FP, Delft 3D) have been developed to address complex real-world hydraulic problems including flood forecasting (Gilles and Moore, 2010; Domeneghetti et al., 2012; Tri et al., 2012). Most of these have developed in a direction of more and more complex and sophisticated modelling approaches in order to achieve highly precise and (often) spatially-explicit, results. In addition, many of these models are complex models that are proprietary and commercial. These factors limit their appeal for wide-spread use in flood forecasting. Moreover, the way in which simulation models are applied has undergone a paradigm shift in recent years: traditionally these were strictly limited to the domain of modelling specialists. However, with the emergence of wider-stakeholder engagement in co-learning, co-designing and co-solving of problems (Akpo et al., 2015), there is now a need for non-restrictively licensed (e.g., open source), simple models for the wide-spread deployment in contemporary co-learning environments.

Probabilistic flood forecasting requires flood models that are simple and fast. Taking into account the large uncertainties in future forcings and different model parameters, a high degree of accuracy of models, are sought after in deterministic applications. They often become secondary to their simplicity in use and rapidity in execution in probabilistic applications. This demands models that are fit-for-purpose by being fast in execution and simple to be deployed in iterative contexts to realize thousands of models runs.

This chapter focuses on the development of such a 'fit-for-purpose' modelling approach suitable for probabilistic flood forecasting which also accounts for forcing and system uncertainty to forecast flood at Can Tho city in the Mekong Delta, Vietnam.

The contents of this chapter are structured as follows: Section 2.2 provides the research methodology. Results obtained and discussion are presented in Section 2.3. The final section presents the main conclusions of this chapter.

2.2 Methodology

2.2.1 Data

The data for this study were collected from two sources: (i) upstream flow (discharge) from 2000 to 2006, measured water level in 2000 at Chau Doc, Tan Chau, Can Tho, Tran De, Ben Trai and An Thuan stations (Figure 1.1. b in Chapter 1), cross-section data and the Manning's roughness coefficient of links were taken from the aforementioned Mekong Delta 1D ISIS model of the MRC. Additionally, discharge in 2011 was collected from MRC as well; (ii) measured water level of the years 2001, 2002 and 2011 at six above gauging stations were collected from the National Hydro-meteorological Service of Viet Nam (NHMS).

Table 2.1. Data sources

Data Type	Source	Data Description
Discharge	MRC	Daily discharge data at upstream (2000–2006) and 2011
Cross-section	MRC	3036 cross-sections representing 8619 km system of river and canals in the Mekong Delta
Manning's roughness	MRC	10 different values of Manning's roughness coefficient
Measured water level	MRC	Hourly water level data at Chau Doc, Tan Chau, Can Tho, Tran De, Ben Trai and An Thuan stations in 2000
	NHMS	Hourly water level data at Chau Doc, Tan Chau, Can Tho, Tran De, Ben Trai and An Thuan stations in 2001, 2002, and 2011

2.2.2 Model selection

To circumnavigate the complexities arising due to simulation time and proprietary nature of sophisticated process-based models such as those used in the studies summarized in Section 1.3.2 of Chapter 1, here, to achieve the objective of this study, the open-source SWMM model was selected. SWMM is a dynamic rainfall-runoff simulation model used for single event or long-term (continuous) simulation of runoff quantity and quality (Rossman, 2015).

The application of the SWMM model for this task was unconventional. This decision was made based on a number of considerations. Firstly, SWMM, solving the one-dimensional Saint Venant flow equations in conduits (Rossman, 2015), is technically capable of simulating the required flow conditions in this complex river system. It can manage complex flow conditions such as downstream forcing by tide. Secondly, it provides a simple, uncluttered user interface that is simple enough to be deployed in multi-stakeholder co-design sessions. Finally, it is a public-domain, open-source model which is not bound by restrictions associated with commercial licenses. Although commercial licensing is not always a significant barrier for a model to be used by experts in, for example, a consulting environment, delivering a model for wider use by non-expert stakeholders is sometimes severely hampered by such restrictions

The flow in natural rivers comprises 1D unsteady flow, which is represented by the Saint-Venant equations (Equations 1.7 and 1.8 in Chapter 1). This governing equation is solved via dynamic wave flow routing in the SWMM model. Natural rivers can be simulated in the SWMM model by a network of open conduits, which are connected by junction nodes. Flooding occurs in the system when the water depth at a node exceeds the maximum depth at node (i.e. the distance from the invert to the ground surface), and the excess flow is lost from the system or can pond atop the node and re-enter the system when the flow in conduits reduces (Rossman, 2015), mimicking what occurs in natural rivers. When the water level in the river is higher than the bank elevation, the flow will overflow and enter the floodplain along the river. The lost flow may return to the system when the water level in the river drops again. In this case study, the width and the bank elevation of Mekong River cross-sections are large and high enough to convey the flow without losing flow from the system.

2.2.3 Model application

SWMM Model Development and Simplification

In the first step of this study, a very detailed SWMM model of the entire Mekong Delta, comprising 575 nodes and 592 conduits, was developed (Fig 2.1). This model covered only the important tributaries of the Mekong, ignoring small canals and other waterways.

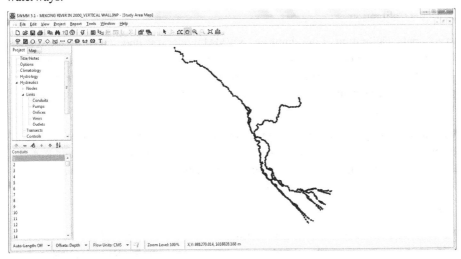

Figure 2.1. The detailed SWMM model for Mekong Delta with 575 nodes and 592 conduits

The input data used were the daily upstream flow and hourly downstream sea level in 2000. This model is already simplified compared to many of the models described in the section above and took about 10 min for a one year simulation (for details see Table 2.10). The next step is to reduce the model complexity such that very fast individual simulations are possible. To this end, a series of simulations (hourly temporal resolution) of the detailed SWMM model were undertaken where the level of detail in the model was gradually decreased (i.e., systematic removal of small to medium size tributaries, based on their size and/or distance from target area). In doing this here nodes that appeared to be at non-critical locations were sequentially removed from the system while retaining nodes at critical locations where there were river branch divisions, changes in the direction of flow, significant changes in cross-sectional area, hydrological stations, etc. The model obtained after each stage in the reductions process was run for the year 2000, and results compared with measured water levels at Chau Doc, Tan Chau and Can Tho stations in 2000. If the post-reduction model results were still good enough, the process of reduction was continued until the simplest level of detail that provides sufficiently accurate predictions of water levels in the local study area (Can Tho, Vietnam) was obtained. The final model thus obtained is referred to herein as the simplified SWMM model. The Simplified SWMM model took only a minute to complete one year period of simulation, representing a 10 and 90 fold reduction in run time relative to the detailed SWMM model and the detailed ISIS model described above (see also Table 2.10).

Model Calibration and Validation

Model Calibration

An automatic calibration of the simplified SWMM model was undertaken for the year 2000 using observed hourly water levels at Chau Doc, Tan Chau and Can Tho stations. The SWMM5-EA software (Pathirana, 2014), which uses evolutionary algorithms to optimize drainage networks, was used to optimise the Manning's roughness coefficient of conduits of the simplified SWMM model.

Water levels at the gauging stations in the Mekong Delta sharply change between flood season and dry season. Therefore, monthly NSE (Nash-Sutcliffe efficiency) and RMAE (Relative Mean Absolute Error) values of the 3 gauging stations Can Tho, Chau Doc and Tan Chau were used for model calibration.

- NSE indicator

The NSE is a normalized statistic that determines the relative magnitude of the residual variance ("noise") compared to the observed data variance ("information") (Nash and Sutcliffe, 1970). Model performance is commonly classified for various ranges of the NSE as shown below (Nash and Sutcliffe, 1970).

$0.75 < NSE \leq 1.00$: Very Good;

$0.65 < NSE \leq 0.75$: Good;

$0.50 < NSE \leq 0.65$: Satisfactory;

$NSE \leq 0.50$: Unsatisfactory.

- RMAE indicator

The RMAE is a commonly used error statistic (Sutherland et al., 2004).

Model performance is commonly classified for various ranges of the RMAE as shown below (Sutherland et al., 2004).

RMAE < 0.20: Excellent;

$0.20 \leq$ RMAE < 0.40: Good;

$0.40 \leq$ RMAE < 0.70: Reasonable;

$0.70 \leq$ RMAE ≤ 1.00: Poor;

RMAE > 1.00: Bad.

Model Validation

Model validation was performed manually for the years 2001, 2002 and 2011, using the model calibrated with the year 2000 data, to gain confidence in model predictions. Results of the validation simulations were compared with hourly water level data acquired at Chau Doc, Tan Chau and Can Tho stations.

2.3 Results and Discussion

2.3.1 Calibration of the Simplified SWMM Model

After a number of trials with varying levels of complexity, a simplified model where the number of nodes and conduits were 37 and 40 respectively (Fig 2.2), was determined to have a positive result (see Tables 2.3 and 2.4) while being sufficiently fast (1 min run time for a 1 year simulation period at hourly temporal resolution). Figures 2.3 and 2.4 show the comparison between simulated water levels with observed water levels at Can Tho station in 2000.

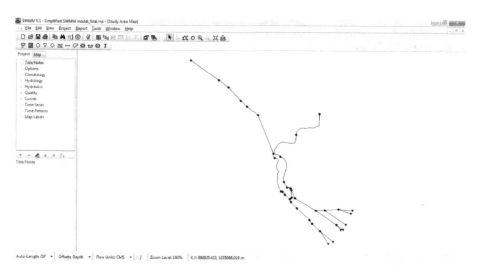

Figure 2.2. The simplified model for Mekong delta with 37 nodes and 40 conduits

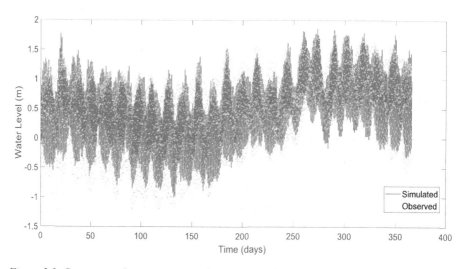

Figure 2.3. Comparison between simulated and observed water level (relative to MSL) time series at Can Tho station in 2000

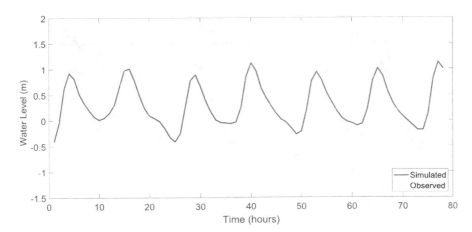

*Figure 2.4. Zoom-in of the comparison between simulated and observed water levels (relative to
MSL) at Can Tho station from 07/03/2000 to 10/03/2000*

The results of the simplified SWMM calibration for the year 2000 is shown in Table
2.2. Scatter plots of simulated and observed hourly water level of each month in 2000 at
Chau Doc, Tan Chau and Can Tho stations are shown in Figs 2.5–2.7, respectively.

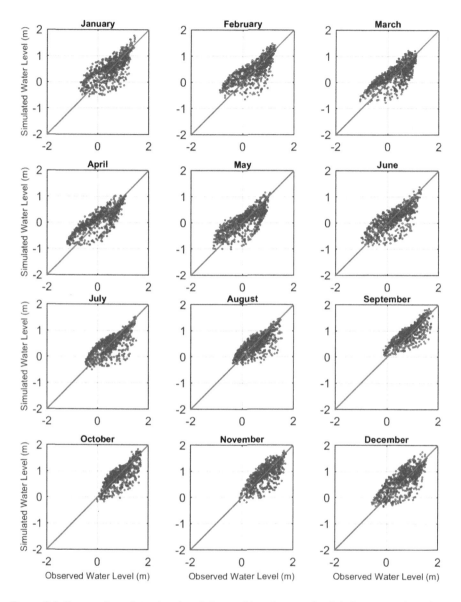

Figure 2.5. Scatter plots of simulated and observed hourly water level (relative to MSL) of each month in 2000 at Can Tho station

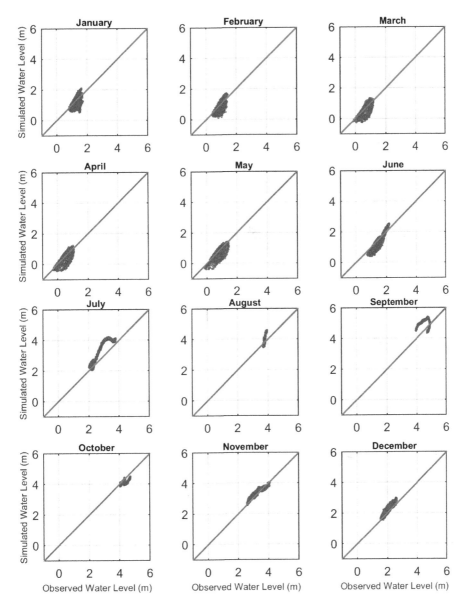

*Figure 2.6. Scatter plots of simulated and observed hourly water level (relative to MSL) of each
month in 2000 at Chau Doc station*

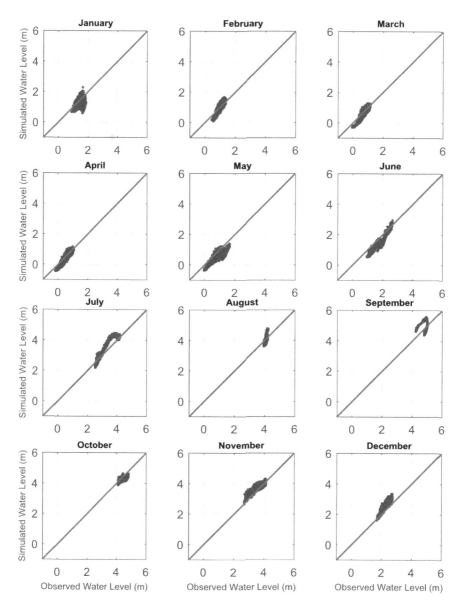

Figure 2.7. Scatter plots of simulated and observed hourly water level (relative to MSL) of each month in 2000 at Tan Chau station

*Table 2.2. General performance rating for the model for the calibration period (year 2000)
using the NSE and RMAE indicators (hourly resolution)*

Station / Month	Chau Doc		Tan Chau		Can Tho	
	NSE	RMAE	NSE	RMAE	NSE	RMAE
January	-0.89	0.05	-0.89	0.12	0.67	0.04
February	-0.14	0.01	0.16	0.09	0.71	0.12
March	0.02	0.07	0.37	0.13	0.72	0.10
April	0.04	0.16	0.51	0.21	0.75	0.80
May	0.27	0.19	0.34	0.28	0.74	0.17
June	0.54	0.07	0.41	0.16	0.68	0.40
July	0.03	0.16	0.76	0.05	0.71	0.09
August	-16.78	0.01	-8.54	0.02	0.72	0.10
September	-2.41	0.09	-3.15	0.07	0.81	0.08
October	-1.05	0.04	-0.92	0.03	0.76	0.08
November	0.77	0.04	0.59	0.05	0.71	0.04
December	-0.53	0.13	-1.17	0.14	0.68	0.05

Based on the results in Table 2.2 and the classifications for NSE as well as RMAE, error
classification for NSE and RMAE indicators for each station in 2000 are shown in
Tables 2.3 and 2.4 respectively.

Table 2.3. Error classification (%) for NSE indicator

Station	Classification for NSE			
	Very Good	Good	Satisfactory	Unsatisfactory
Chau Doc	8.33	0.00	8.33	83.33
Tan Chau	8.33	0.00	16.67	75.00
Can Tho	16.67	83.33	0.00	0.00

Table 2.4. Error classification (%) for RMAE indicator

Station	Classification for RMAE				
	Excellent	Good	Reasonable	Poor	Bad
Chau Doc	100.00	0.00	0.00	0.00	0.00
Tan Chau	83.33	16.67	0.00	0.00	0.00
Can Tho	83.33	0.00	8.33	8.33	0.00

Based on the results in Tables 2.3 and 2.4, almost RMAE values at Chau Doc, Tan Chau and Can Tho stations are in excellent and good classifications, especially Chau Doc station with 100 % values is in excellent classification, while Can Tho is 83.33 %. Nonetheless, Can Tho also has 8.33 % values in total is poor classification. For NSE indicator, only Can Tho station has classifications are in very good and good with 100 % in total, while Chau Doc and Tan Chau are 8.33 %. The main classifications of Chau Doc and Tan Chau stations are unsatisfactory with 83.33 % and 75 %, respectively.

2.3.2 Model validation

The results of the S-SWMM validation for 2001, 2002 and 2011 are shown in Tables 2.5–2.7, respectively.

Table 2.5. General performance rating for the model for the validation period (year 2001) using the NSE and RMAE indicators (hourly resolution)

Station / Month	Chau Doc		Tan Chau		Can Tho	
	NSE	RMAE	NSE	RMAE	NSE	RMAE
January	-1.08	0.22	-0.05	0.13	0.66	0.21
February	0.85	0.32	0.95	0.18	0.79	0.21
March	0.12	0.06	0.64	0.00	0.75	0.11
April	0.19	0.01	0.62	0.02	0.78	0.23
May	0.14	0.08	0.49	0.12	0.76	5.98
June	0.01	0.20	-0.22	0.28	0.73	0.52
July	0.56	0.00	0.05	0.07	0.68	0.29
August	0.57	0.09	0.88	0.01	0.67	0.06

September	-12.15	0.01	-13.35	0.04	0.79	0.06
October	0.76	0.00	0.31	0.03	0.63	0.06
November	0.37	0.07	0.61	0.03	0.71	0.03
December	-0.52	0.17	-0.05	0.12	0.68	0.05

Table 2.6. General performance rating for the model for the validation period (year 2002) using the NSE and RMAE indicators (hourly resolution)

Station / Month	Chau Doc		Tan Chau		Can Tho	
	NSE	RMAE	NSE	RMAE	NSE	RMAE
January	-2.00	0.30	-0.33	0.16	0.66	0.16
February	0.83	0.29	0.96	0.14	0.68	0.19
March	0.15	0.17	0.76	0.01	0.75	0.22
April	-0.21	0.08	0.42	0.10	0.67	0.28
May	0.28	0.01	0.54	0.17	0.77	0.24
June	0.46	0.13	-0.02	0.28	0.74	0.15
July	0.70	0.05	0.67	0.09	0.65	0.41
August	0.46	0.12	0.89	0.01	0.74	0.14
September	-3.49	0.03	-0.81	0.02	0.75	0.13
October	0.69	0.03	0.70	0.01	0.75	0.08
November	-1.57	0.17	-0.01	0.09	0.70	0.01
December	-1.83	0.23	-0.53	0.15	0.66	0.04

Table 2.7. General performance rating for the model for the validation period (2011) using the NSE and RMAE indicators (hourly resolution)

Station / Month	Chau Doc NSE	Chau Doc RMAE	Tan Chau NSE	Tan Chau RMAE	Can Tho NSE	Can Tho RMAE
January	-4.76	0.80	-5.34	0.74	0.45	0.54
February	0.47	0.67	0.63	0.63	0.72	0.60
March	-0.50	0.38	0.03	0.37	0.64	0.53
April	-0.10	0.29	0.51	0.27	0.67	0.86
May	0.03	0.29	0.62	0.21	0.68	1.79
June	-0.16	0.19	0.68	0.01	0.69	1.29
July	-4.17	0.33	-0.03	0.09	0.72	0.30
August	-3.61	0.36	0.11	0.11	0.75	0.12
September	-5.96	0.23	-0.06	0.08	0.78	0.10
October	-17.46	0.07	-0.41	0.02	0.77	0.04
November	-1.57	0.20	-0.06	0.11	0.64	0.10
December	-6.38	0.45	-3.80	0.35	0.58	0.23

Based on the results in Tables 2.5–2.7 and the classifications for NSE and RMAE, error classification for NSE and RMAE indicators for each station in 2001, 2002 and 2011 are shown in Tables 2.8 and 2.9 respectively.

Table 2.8. Error classification (%) for NSE indicator in 2001, 2002 and 2011

Station	Classification for NSE			
	Very Good	Good	Satisfactory	Unsatisfactory
Chau Doc	8.33	5.56	5.56	80.55
Tan Chau	13.89	8.33	19.44	58.33
Can Tho	19.44	63.89	13.89	2.78

Table 2.9. Error classification (%) for RMAE indicator in 2001, 2002 and 2011

Station	Classification for RMAE				
	Excellent	Good	Reasonable	Poor	Bad
Chau Doc	55.56	36.11	5.56	2.78	0.00
Tan Chau	77.78	16.67	2.78	2.78	0.00
Can Tho	50.00	25.00	13.89	2.78	8.33

Tables 2.8 and 2.9 show that a vast majority of RMAE values at Chau Doc, Tan Chau and Can Tho stations are in excellent and good classifications. Specifically, the percentages of data/model comparison that fall into excellent and good categories are 91.67 %, 94.45 % and 75 % at Chau Doc, Tan Chau and Can Tho, respectively. However, it is noted that 2.78 % of RMAE values fall into the poor category for all 3 stations, while at Can Tho 8.33 % RMAE values also fall into the bad category. With respect the NSE indicator, 83 %, 13.89 % and 22.22 % of NSE values fall in very good and good categories (in combination) at Can Tho, Chau Doc, and Tan Chau stations, respectively. The majority of NSE values at Chau Doc (80.55 %) and Tan Chau (58.33 %) are unsatisfactory.

For the RMAE indicator, the poor and bad classifications are generally in the dry months (from December to June), while the good and excellent categories are generally in the wet months (from July to November). However, it is noteworthy that at Can Tho, the target area pertaining to this study, the differences between the simulated monthly mean water levels and the measured monthly mean water levels are small even during wet months, the maximum value is 0.15 m. As the main focus of this study is flooding during wet months at Can Tho, the weaker data/model comparison, especially during dry months, at the secondary stations of Chau Doc and Tan Chau do not have major implications on achieving study objectives.

For the NSE indicator, unsatisfactory data/model comparisons are shown mainly at Chau Doc and Tan Chau stations, while at Can Tho NSE is unsatisfactory on very few occasions (2.78 %), and that too during dry months only. While it is possible to improve the NSE values for the two secondary stations by adding nodes and conduits to the simplified model, that will invariably increase the runtime for each simulation, which detracts from the main purpose of this study and will not add any great value to our main objective of developing a model that is able to provide probabilistic estimates of flooding at Can Tho.

Thus, for the purpose of obtaining accurate and fast predictions of flood water levels at Can Tho, the simplified SWMM model performance can be considered good enough.

2.3.3 Performance comparison between the previous models and the simplified SWMM model for the entire Mekong Delta

The characteristics and simulation time associated with the process based hydrodynamic models used in previous studies of the Mekong Delta (Section 1.3.2 of Chapter 1) and the simplified SWMM model used here are shown in Table 2.10.

Table 2.10. Characteristics and simulation time of different models for the entire Mekong Delta (for a single core on a Intel(R) Core(TM) i5-4210M CPU @ 2.6GHz 2.6GHz processor in a computer with 8.0 GB of memory (RAM))

Model Name	Number of Nodes and Links		Simulation Time for one Year (minutes)
	Node/Junction	Link/Cross-Section	
ISIS model (1D)	572	3,036	90
VRSAP model	1,505	2,111	–
Hydro-GIS model	–	13,262	–
Detailed SWMM model	575	592	10
Simplified SWMM model	37	40	1

2.4 Conclusions

This chapter demonstrated that it is possible to simulate river water levels, with an acceptable level of accuracy (good – excellent) (i.e. the model is sufficiently reliable for flood forecasting), at a location of interest in a complex, deltaic river system like the lower Mekong with a relatively simple (simplified from thousands to several tens of cross-sections) and fast (reduction of simulation time from 1.5 h to around one minute, for a 1 year simulation) numerical model. The simplified model was achieved by iteratively simplifying a complex hydraulic model, with the focus on the accuracy of the water levels at a single point of interest (in this case, near Can Tho city).

The objective of developing this simplified and fast model was to subsequently use it for probabilistic flood simulations and for stakeholder-based co-design applications (via a co-design approach to identify appropriate risk reduction measures with effective stakeholder involvement). Probabilistic flood modelling involves running thousands of

realizations of the model for a given scenario. Even with advanced computer facilities
available today, this requirement makes it prohibitive to use some of the 'traditional'
models that take hours to run one simulation. Towards this end, the simplified SWMM
model presented in this chapter provides a feasible solution. The computing cost is still
considerable (e.g., thousand simulations will cost around 17 core-h), but with a modern
computer with sixteen cores and adequate amount of memory, such a simulation should
be completed within a little more than an hour.

The application of the SWMM, a model developed to simulate drainage/sewerage
systems, has been shown to be capable of simulating river systems with complex
boundary conditions with positive results. Additionally, with the advantage of being an
open-source model and a simple user interface, it would be an appropriate option for
multi-stakeholder co-design meeting. Furthermore, the SWMM model also provides a
well-documented, clear application programming interface (API) making it possible to
embed the model in computer software applications written in a variety of computer
languages like C/C++, Fortran and Python. This opens up the possibility of using the
resulting model as a basis for innovative applications such as serious-gaming.

The simplification of a model can sometimes lead to a degradation of the precision and
accuracy of its results. The appropriateness of a simplified model should be looked at in
the context of the ultimate intended use of model outcomes. The intended application of
the simplified SWMM model for the Mekong delta is two-fold: (a) to derive
probabilistic river water level estimates to be used as input for a probabilistic urban
flood model; and (b) to be used as a co-design tool in multi-stakeholder environments.
In terms of both these utilities, the level of uncertainty associated with the input
information is significant. For example, future sea level projections and upstream flow
conditions have large uncertainties associated with them (more in latter than in the
former). Therefore, striving for high precision modelling output under these realities
does not improve the accuracy of the ultimate intended outcomes. In such
circumstances, it is appropriate to sacrifice some accuracy to achieve efficiency and
practicality.

Chapter 3

Developing probabilistic flood hazard maps for the urban centre of Can Tho city for present-day and future[2]

---

[2] This chapter is partially based on Ngo, H., Ranasinghe, R., Zevenbergen, C., Kirezci, E., Maheng, D., Radhakrishnan, M. and Pathirana, A.: An Efficient Modelling Approach for Probabilistic Assessment of present-day and Future Fluvial Flooding, Frontiers in Climate, submitted, 2021.

3.1 Introduction

Flooding is one of the most frequently occurring and damaging natural disasters in the world (Hirabayashi et al., 2013; Kundzewicz et al., 2014; Arnell and Gosling, 2016; Alfieri et al., 2017; Forzieri et al., 2017; Mora et al., 2018). Flood hazard estimation, which computes the probability and intensity of a possible event (Pappenberger et al., 2012; Alfieri et al., 2013; De Moel et al., 2015) is the first step in flood risk assessment (Penning-Rowsell et al., 2005; De Moel et al., 2015; Foudi et al., 2015; Kvočka et al., 2016). Flood hazard maps obtained from flood hazard assessment are used for estimating the potential socio-economic impacts of flooding (Koks et al., 2015). FRM and planning decisions in many parts of the world have historically utilised flood hazard or risk maps associated with one or two pre-determined return period of water level (e.g. 100-year, 200-year) (Merz et al., 2007). However, currently, the need to move from stationary to innovative time-dependent (non-stationary) flood hazard and risk modelling approaches that can account for the uncertainty, arising from anthropogenic and climatic induced stressors, is rapidly increasing (Mosavi et al., 2018).

Climate change is recognized as a major global challenge in the 21st century and beyond, which may lead to increased flooding in the future (Panagoulia and Dimou, 1997; Menzel et al., 2002; Prudhomme et al., 2013; Alfieri et al., 2015). Besides the challenges posed by climate change, population increase inevitably increases water demand, which is often satisfied by excessive groundwater extraction, which, more often than not, leads to land subsidence, further exacerbating the flood hazard (due to increased inundation levels). The combination of increased flood hazards and increased population/infrastructure will, in turn, lead to an increase in flood risk in vulnerable areas (Nicholls et al., 2007; Lenderink and Van Meijgaard, 2008; Syvitski et al., 2009; Min et al., 2011; Balica et al., 2012; Rojas et al., 2013; Wong et al., 2014). However, projections of these trends contain uncertainties. This particularly holds for deltas due to their multi-faceted and dynamic character (Nicholls et al., 2020).

Robust quantification of flood risk typically requires multiple (thousands of) river and flood model simulations to derive probabilistic flood hazard distributions. This constitutes a major challenge, especially for large river systems, such as the Mekong, due to the associated computational and resource demands. Although the last decade has witnessed a great improvement in computational capabilities and models, traditional modelling approaches still pose significant challenges in terms of computational time required to obtain fully probabilistic flood hazard estimates (Mcmillan and Brasington. 2007; Neal et al., 2012). Therefore developing computationally efficient modelling approaches to circumvent this particular bottleneck remains an important challenge.

The chapter attempts to address the challenge of estimating non-stationary fluvial flood hazard in a way that it can inform flood risk modelling via a computationally efficient modelling approach. The approached adopted here is based on a simplified 1D hydrodynamic model for the entire Mekong Delta (area of 40,577 km^2) that is coupled with a detailed 1D/2D coupled model, and here this approach is demonstrated at Can Tho city in the Mekong Delta.

This chapter also presents probabilistic flood hazard maps for the urban centre of Can Tho city (Ninh Kieu district) for present-day and future under different scenarios taking into account the impact of climate change forcing (river flow, sea-level rise, storm surge) and land subsidence.

The contents of this chapter are structured as follows: Section 3.2 provides a brief description of the study area. Research methodology is presented in Section 3.3. Section 3.4 presents results and discussion. Conclusions of this chapter is presented in Section 3.5.

3.2 Study area

This chapter focuses on the part of the Ninh Kieu district (Fig. 3.1c), which is the most developed and major urban district in Can Tho. It is home to many trade centres, urban areas, and residential areas. Ninh Kieu has the highest population among the five inner districts, with a population of approximately two hundred eighty thousand over an area of twenty-nine square kilometre as of 2019 (population density approx 9,600 per km^2). It is also the most severely flooded inner district in Can Tho City with an inundation depth of up to 70-80 cm on some main roads in the previous flood events (CCCO & ISET, 2015).

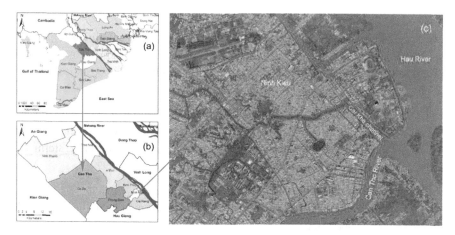

Figure 3.1. (a) Map of Mekong delta and surrounding provinces, (b) Can Tho city, (c) Study area (bounded by red line) within the Ninh Kieu district in Can Tho city (Base map is from Bing Maps satellite © Microsoft)

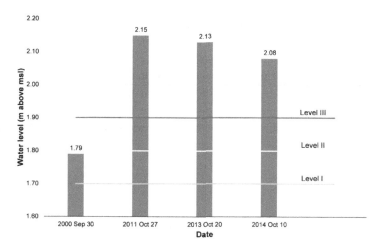

Figure 3.2. Highest measured water levels of the flooding years since 2000 and flood water level alarms in Can Tho (flood water level alarms are implemented following Decision No.632/QĐ-TTg issued on May 10th, 2010)

3.3 Methodology

To achieve the goal of producing probabilistic flood hazard estimates for the study area (for the present and 2050), the method adopted here broadly comprised three methodological steps: (a) simulation of a large number (over 1000) of upstream boundary conditions with a fast 1D hydrodynamic model for the entire Mekong Delta, developed and validated in Section 2.3.2 of Chapter 2, (b) reduction of the number of 1D/2D coupled model runs required by flood frequency analysis, and (c) derivation of probabilistic flood hazard quantification for the present and for 2050 under RCP 4.5 and 8.5, with and without land subsidence. The methodology adopted is summarized in Fig. 3.3 and described below.

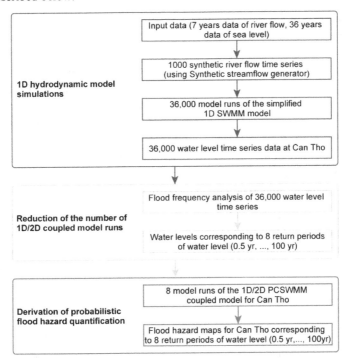

Figure 3.3. The methodological framework adopted to derive probabilistic flood hazard estimates for the study area. This process was applied here for the present and for 2050 (RCP 4.5 and RCP 8.5 together with 3 different local land subsidence scenarios)

The current situation and four different scenarios (Table 3.1) were simulated with the 1D/2D flood model.

Table 3.1. Model scenarios adopted with the 1D/2D coupled flood model

Model scenario	Time	Climate scenario	Land subsidence rate
1	2050	RCP 4.5	1.6 cm/yr
2	2050	RCP 4.5	0

3	2050	RCP 8.5	1.6 cm/yr
4	2050	RCP 8.5	0

3.3.1 1D hydrodynamic model simulations

Upstream boundary condition:

In the application here, seven years (2000-2006) of observed flow data (was used in the detailed ISIS model for the Mekong Delta and established by MRC) at the upper boundary location 'Kratie' were used as forcing for the simplified 1D model of the Mekong River obtained in Chapter 1. As seven years of data is not sufficient to derive probabilistic results, here a synthetic streamflow generator developed by Giuliani et al. (2017) was used to generate 1000 synthetic flow time series (each one year long) for the current situation and each future scenario in Table 3.1. While it is acknowledged that this approach does not fully replace the utility of long term observational data, due to the probabilistic nature of the streamflow generator, it nevertheless captures the statistical variation of upstream flows compared to simply using the seven years of available data, which is important in flood hazard modelling.

For the future simulations that include the effects of climate change, the annual river discharge projections of Hoang et al. (2016) under RCP 4.5 and 8.5 were used as the basis for generating the future river flow data. Hoang et al.'s (2016) projections indicate that annual river flow for RCP 4.5 and RCP 8.5 at Kratie in 2050 are expected to change between 3 % to 8 % and -7 % to 11 %, respectively, relative to the 1971-2000 baseline period adopted in that study. Future river flow data were generated by combining the flow of each year in seven years of observed flows used here with a randomly selected % change of these projected changes in the flow corresponding to each RCP. Subsequently, the streamflow generator was used to generate 1000 future riverflow time series (each one year long) corresponding to RCP 4.5 and RCP 8.5.

Downstream boundary condition:

Thirty six years (1979-2014) of model simulated extreme sea level (tide + surge) data were extracted at the Mekong River mouths (Tran De, Ben Trai and An Thuan) from the GTSR data set presented by Muis et al. (2016) to use as the downstream boundary condition of the 1D model. This is the only available extreme sea level hindcast for the study area.

For the future simulations, the above present-day extreme sea levels from the GTSR data were combined with the 2050 regional sea level rise projections presented by the Viet Nam Ministry of Nature Resource and Environment (MONRE, 2016) under RCP 4.5 and RCP 8.5 for the region containing the aforementioned river mouths. The projected regional sea level rise in the area by 2050 (relative to 1986-2005) for RCP 4.5 and RCP 8.5 are 13 to 32 cm and 16 to 35 cm, respectively.

The simplified 1D model was then executed with one year time series of the above described boundary conditions to generate river water level at Can Tho. In all, 36,000 simulations were undertaken corresponding with the total number of possible combinations of upstream (1000 one-year time series of riverflow) and downstream (36 one-year time series of sea level) boundary conditions, resulting in 36,000 water level time series data (each one-year long) at Can Tho per considered scenario (current and future).

3.3.2 Reduction of the number of 1D/2D model runs

From the 36,000 one-year long water level time series at Can Tho, the maximum water level of each year in the 36,000 simulation years was extracted and used to fit an extreme value distribution (Gumbel or Type I). The water level at Can Tho corresponding to each return period for each scenario (current or future) was determined based on the fitted Gumbel distribution (Gumbel, 1935).

Apart from the maximum water levels, it is also important in flood hazard modelling to represent the shape of the hydrograph around the peak water level. To achieve this, a threshold-based method has been used, which is commonly used in flood frequency analysis (Lang et al., 1999; Bezak et al., 2014). A threshold value of 2.15m (i.e. maximum measured water level at Can Tho, see Fig. 3.2) was adopted here and annual water level time series (of the full 36,000 series) that contain at least one water level value exceeding 2.15m were identified. From each of the extreme water level time series corresponding to RCP 4.5 and 8.5 thus identified (41 and 162 for RCP 4.5 and RCP 8.5 respectively), 24 h long time series around each peak value (12 h earlier to 12 h later) were extracted.

3.3.3 Derivation of probabilistic flood hazard quantification

Several different flood parameters can be used to quantify the flood hazard, including inundation level, flow velocity, frequency of flooding, and flood duration, etc. (Ramsbottom et al., 2003; Ward et al., 2011a; Moel et al., 2015). Of these, inundation level (water depth) and flow velocity are considered the most important parameters (Penning-Rowsell et al., 1994; Wind et al., 1999; Merz et al., 2007; Kreibich et al., 2009). However, due to the relatively flat terrain combined with small inundation depths in Can Tho, the effect of the flow velocity is expected to be small compared to that of the flood inundation depth (Dinh et al., 2012). Hence, this study considers inundation levels as the main indicator of the flood hazard in the study area.

To simulate the effect of the flood drainage network on flooding in the study area, the detailed 1D urban model for the study area are developed by Huong and Pathirana (2013), has been used. This model comprises 479 junctions, 612 conduits, 48 outfalls, and 303 sub-catchments (Fig. 3.2).

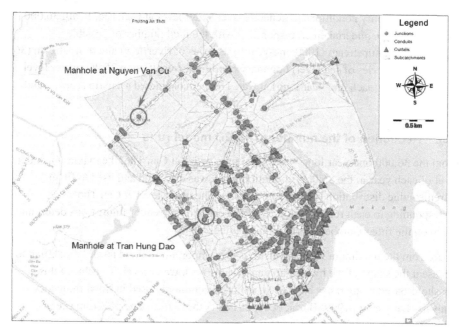

Figure 3.4. The detailed 1D urban flood drainage model for the study area developed by Huong and Pathirana, (2013) (Base map is from Bing Maps Road © Microsoft)

The model parameters were calibrated for the flood event on the 17th of October 2016 based on the observed water depths (at one-minute measurement interval) in the manholes at Nguyen Van Cu and Tran Hung Dao streets (Fig. 3.4). These are the only available observed flood water depths in the study area. Here, SWMM5-EA software (Pathirana, 2014), was used to calibrate the most uncertain parameters, i.e. Manning's roughness coefficient of conduits, Manning's roughness coefficient of the previous/impervious surfaces of the sub-catchments and the slope of the sub-catchments. The calibrated 1D urban model was then used to establish an integrated 1D/2D model using PCSWMM software (http://www.chiwater.com/Software/PCSWMM). The ability of the 1D/2D model to simulate the spatial pattern of flooding was verified by simulating the 17th October 2016 flood event.

Flood simulations were then undertaken for water levels with return periods ranging
from 0.5 yr to 100 yr obtained from the flood frequency analysis described in Section
3.3.2 for each model scenario in Table 3.1. A 15 m resolution DEM developed by the
Vietnam Institute of Meteorology, Hydrology and Environment (Huong and Pathirana,
2013), was used in all simulations. In the simulations that include land subsidence, an
average subsidence rate of 1.6 cm/yr for the entire Mekong Delta (Erban et al., 2014;
Minderhoud et al., 2015) was used to consider the effect of land subsidence on the flood
hazard at Ninh Kieu district.

For all model scenarios indicated in Table 3.1, flood inundation maps were first
developed for water level return periods ranging from 0.5 yr to 100 yr (eight inundation
maps for each model scenario). The flood maps thus obtained were added into ArcMap
under shapefiles, which were then converted to raster files to extract the maximum
inundation depths (during the flood events) corresponding to each return period by
using the "Polygon to Raster" tool in Conversion tools of ArcToolbox. Subsequently,
maximum inundation depths at all grid cells were aggregated to generate flood hazard
maps for each return period.

3.4 Results

Figure 3.5 shows flood frequency curves at Can Tho for the present, and for 2050
corresponding to RCP 4.5 and RCP 8.5, obtained from the Gumbel distributions that
were fitted to the modelled maximum water level at Can Tho based on the results of 1D
hydrodynamic model described in Section 3.3.1.

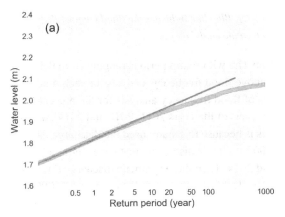

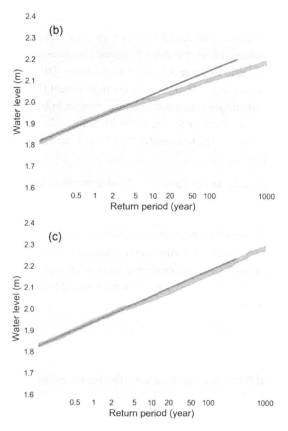

Figure 3.5. Flood frequency curves at Can Tho for Present (a), and for 2050 corresponding RCP 4.5 (b) and RCP 8.5 (c). Blue line indicates the fitted Gumbel distribution while the green line indicates 1D hydrodynamic model output

The water levels at Can Tho with return periods ranging from 0.5 yr to 100 yr determined from the above flood frequency analysis for each model scenario are shown in Fig. 3.6. The results of flood frequency analysis for the present (Fig 3.5 a) indicate that the highest water levels of the years 2011, 2013 and 2014 (see Fig 3.2) are outside of the1000 yr RP. This is because this study used the discharge data at Kratie from 2000 to 2006 to generate 1000 synthetic discharge time series, not including the discharge data of 2011, 2013 and 2014. Then these synthetic discharge time series were combined with 36-year extreme sea levels from the GTSR to generate 36000 water level time series at Can Tho. Therefore, the flood frequency curve for the present based on seven years of discharge (2000-2007) did not cover the water levels that are the combinations of the discharge in 2011, 2013, and 2014 with sea levels. In fact, the 2011-2014 period contains several years in which the maximum annual discharges were uncharacteristically high during hat was otherwise a period (2010-2018) which shows a declining trend in maximum annual discharge (see Figure 3.18). In addition, the accuracy of sea level data at Mekong river mouths, which was extracted from the GTSR (not measured data), can lead to the underestimate in the water levels at Can Tho in the analyses.

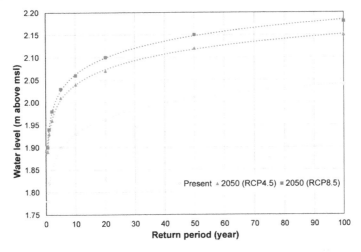

Figure 3.6. The water level at Can Tho corresponding to each return period for the present, and for 2050 under RCP 4.5 and RCP 8.5

The calibration results for the 1D urban drainage model for the flood event of Oct 2016 are shown in Fig. 3.7. The NSE indicator values between the measured and simulated (present conditions) water depths in the manholes at Nguyen Van Cu and Tran Hung Dao streets is 0.75 (good) and 0.95 (very good), respectively. Thus, the 1D urban drainage model performance for the study area can be considered to be sufficiently accurate.

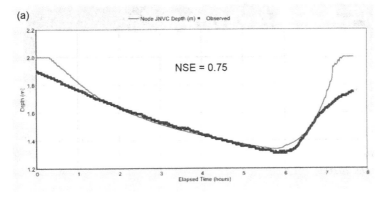

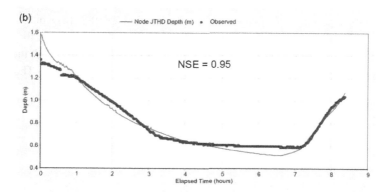

Figure 3.7. Comparison between simulated and observed water depths in the manholes at Nguyen Van Cu (a) and Tran Hung Dao (b) streets on October 17, 2016.

Simulated flood extent and inundation depths corresponding to the flood event of Oct 2016 are shown in Fig. 3.8. Simulated flood depths were compared with measured flood depths in many different streets in the district as reported by Can Tho Water Supply and Drainage Construction Company for this event (BC-XNCTN regarding the situation of high tide October 17, 2016). The comparison shows a good agreement between the simulated flood depths and observations (Fig 3.8)

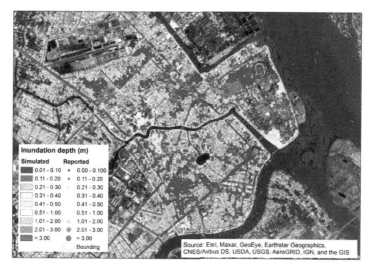

Figure 3.8. Flood inundation map (flood extent and the maximum simulated inundation depth) and the measured inundation depth (dots) at streets in Ninh Kieu district corresponding to the flood event Oct 2016

Figure 3.9 shows the shape of the 24-h time series of all the modelled extreme water level (i.e. peaks greater than the threshold value of 2.15 m) time series at Can Tho (of all 36,000 annual time series) corresponding to RCP 4.5 and RCP 8.5. There are at least two dominant flood hydrograph shapes in the two RCPs considered (indicated by Pattern 1 and Pattern 2 in Fig. 3.9.a and Fig. 3.9.b), but it is clearer in RCP 8.5. Pattern 1 and Pattern 2 are groups of similar water level hydrograph shapes at Can Tho, which were identified after comparing all instances when the water level exceeded a threshold value of 2.15m. These shapes resembles the shape of tides. This is consistent with the fact that the water level and inundation level in Can Tho do depend on the downstream tidal fluctuation (Takagi et al., 2015). The highest 24-h long water level time series of all the water level hydrographs corresponding to Pattern 1 and Pattern 2 were selected as a typical water level hydrograph shape for each pattern respectively. These were scaled to the maximum water level of each event with calculated water level corresponding to 100-year return period to create the 24-h boundary condition time series for the 1D/2D flood model in order to examine the response of different river water level hydrograph shapes on flooding at Ninh Kieu districtThe results (Fig. 3.10) indicates that flood extent and inundation depths associated with Pattern 1 are greater than that associated with Pattern 2. The inundated area (for cells with maximum inundation depths ≥ 0.02 m) corresponding to 100-year return period water level for RCP 4.5 and RCP 8.5 with Pattern 1 are 1.96 km^2 and 2.30 km^2, respectively. The corresponding flooded areas for Pattern 2 are lower at 1.41 km^2 and 1.81 km^2. Therefore, in this study, the typical water level time series following Pattern 1 above was scaled to the maximum water level with calculated water levels corresponding to each return period for each scenario to create the 24-h boundary condition time series for the 1D/2D. This was a necessary simplification to reduce the complexity and computational burden of computing flood inundation.

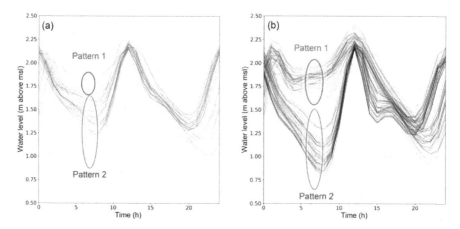

Figure 3.9. Analysis of 24-h time series around modelled peak water levels indicating two flood hydrograph shape patterns in the future scenario, (a) – 2050 (RCP4.5), (b) – 2050 (RCP 8.5)

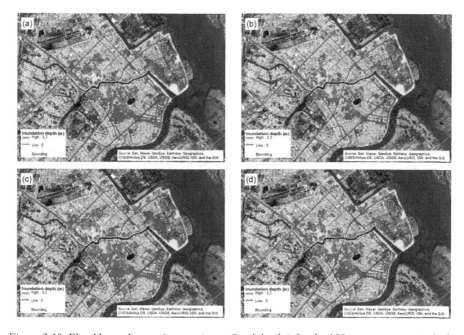

Figure 3.10. Flood hazard maps (i.e. maximum flood depths) for the 100 year return period of water level for hydrograph shape Pattern 1 (left) and Pattern 2 (right) under RCP 4.5 (Fig. 3.10a, Fig. 3.10b) and RCP 8.5 (Fig. 3.10c, Fig. 3.10d) (without land subsidence). The inundated area (for cells with maximum inundation depths ≥ 0.02m) for RCP 4.5 and RCP 8.5 corresponding to Pattern 1 are 1.96 km² and 2.30 km², respectively, while for Pattern 2 are 1.41 km² and 1.81 km², respectively

Figures 3.11-3.13 show flood hazard maps corresponding to 0.5, 5, 50, 100 yr return period water levels (at Can Tho) for the present (model scenario #1), and 2050 under RCP 4.5 and RCP 8.5 (without land subsidence) (model scenarios #3 and #5 respectively). The remaining flood hazard maps which correspond to 1, 2, 10, 20 yr return period water levels for these scenarios are shown in Appendix A (Figs. A1-A3).

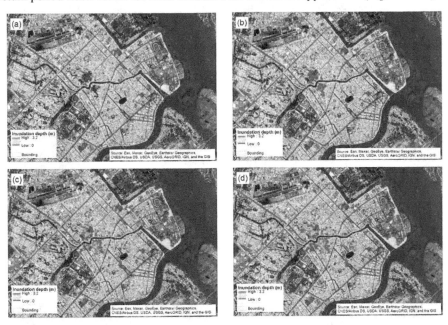

Figure 3.11. Flood hazard maps for the present corresponding to each return period of water level, (a) 0.5 yr return period, (b) 5 yr return period, (c) 50 yr return period, (d) 100 yr return period. Red circle in Fig. 3.11a highlights a large inundated area (present for all return periods) of Cai Khe ward

Figure 3.12. Flood hazard maps for 2050 under RCP 4.5 (model scenario #2) corresponding to each return period of water level, (a) 0.5 yr return period, (b) 5 yr return period, (c) 50 yr return period, (d) 100 yr return period

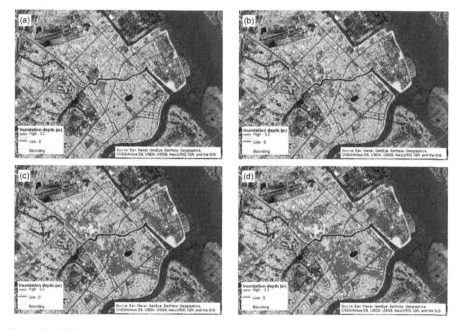

Figure 3.13. Flood hazard maps for 2050 under RCP 8.5 (model scenario #4) corresponding to each return period of water level, (a) 0.5 yr return period, (b) 5 yr return period, (c) 50 yr return period, (d) 100 yr return period

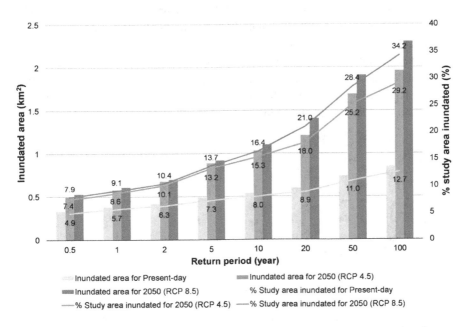

Figure 3.14. The inundated area and percentage of the flooded area relative to total study area (also indicated by numerics along the solid lines in the figure) corresponding to each return period for the present and for 2050 under RCP 4.5 and RCP 8.5 (without land subsidence)

Figures 3.11a and 3.14 show that, under present conditions, 4.9 % of the study area will be flooded even with the 0.5 yr return period of water level at Can Tho, especially some areas along the canals and lakes in the city that are connected to the Can Tho River and Cai Khe channel. Notably, there is a large area of Cai Khe ward (shown by the red circle of Fig. 3.11a), which is located close to the junction of Hau River, Can Tho River and Cai Khe channel (see Fig. 3.1c) that is inundated under this condition.

At the other higher end of the modelled return periods under present conditions, Figs. 3.11d and 3.14 show that the inundated area with a 100 yr return period of water level at Can Tho, is more than double the size of the inundated area with a 0.5 yr return period of water level at Can Tho increasing from 4.9 % (0.5 yr RP) to 12.7 % (100 yr RP), see Fig. 3.14.

For the future modelled scenarios (without land subsidence; model scenarios #2 and #4), the % inundated areas for the 100 yr RP water level at Can Tho are 29.2 % and 34.2 %, under RCP 4.5 and RCP 8.5 respectively, representing more than a doubling of the inundated area by 2050, relative to the present.

Flood hazard maps corresponding to 0.5, 5, 50, 100 yr return period water levels (at Can Tho) for the future scenarios (2050 – under RCP 4.5 and RCP 8.5) that do take into account land subsidence (model scenarios #1 and #3) are shown in Figs. 3.15 and 3.16. The remaining flood hazard maps which correspond to 1, 2, 10, 20 yr return period water levels for these scenarios are shown in the Appendix A (Figs. A4 and A5). The difference is immediately visible with a large increase in the inundated area compared to when land subsidence is not taken into account.

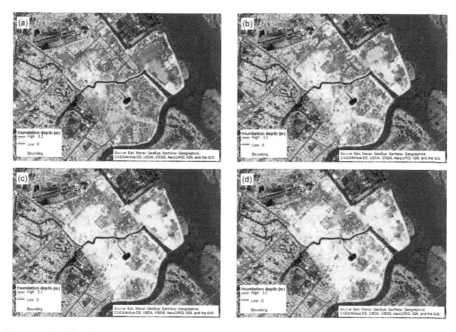

Figure 3.15. Flood hazard maps for 2050 under RCP 4.5 (model scenario #1) corresponding to each return period of water level, (a) 0.5 yr return period, (b) 5 yr return period, (c) 50 yr return period, (d) 100 yr return period

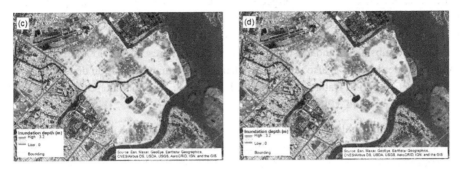

Figure 3.16. Flood hazard maps for 2050 under RCP 8.5 (model scenario #3) corresponding to each return period of water level, (a) 0.5 yr return period, (b) 5 yr return period, (c) 50 yr return period, (d) 100 yr return period

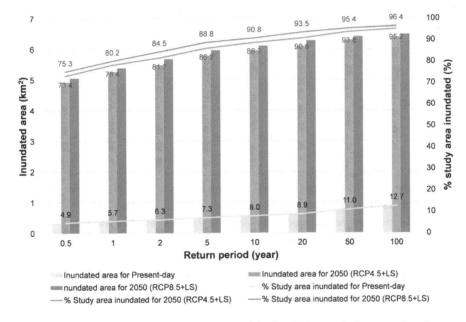

Figure 3.17. The inundated area and percentage of the flooded area relative to total study area (also indicated by numerics along the solid lines in the figure) corresponding to each return period for the present and for 2050 under RCP 4.5 and RCP 8.5 (including land subsidence)

Figures 3.15 and 3.16 both show severe flooding of the study area for all return periods of water level. The percentage area flooded by the 0.5 yr RP water level under RCP 4.5 and RCP 8.5 are 73.4 % and 75.3 %, respectively (Fig. 3.17). This is almost a 10-fold increase relative to the projections without land subsidence and a 15-fold increase relative to the present-day flooding for the same RP water level. For the 100 yr RP water level, the difference in the projections are even more extreme with the percentage area flooded under RCP 4.5 and RCP 8.5 projected to be 95.2 % and 96.4 % respectively, representing a 3-fold increase relative to the comparable projections without land subsidence and a 8-fold increase relative to present-day flooding for the same RP water level. It is also noteworthy that the maximum inundation depths for the 100-year return period water level under RCP 4.5 and 8.5 are not much different: these values are 3.17 m, and 3.20 m, respectively.

The above results highlight that while climate change will increase the flood hazard in the study area, land subsidence has a much greater effect than climate change driven variations in river flow on the flood hazard in the study area. Furthermore, clearly, the existing urban drainage network is not able to effectively drain flood waters even for present-day conditions, and this will be felt more severely in the coming decades. A significant reduction in groundwater extraction, which is the main cause of land subsidence in Can Tho (Erban et al., 2014) combined with a new and substantially efficient urban drainage network may be able to mitigate the projected flood hazard in the study area. It is recommended that the effectiveness of these mitigation measures be investigated in detail in future modelling studies.

3.5 Discussion

This Chapter presented probabilistic fluvial flood hazard maps for the Ninh Kieu district for present-day and future under different scenarios taking into account the impact of climate change forcing (river flow, sea-level rise, storm surge). Although computation of present-day flood hazard and probabilistic flood maps using 2D models for Ninh Kieu district has also been done before (e.g. Apel et al. (2016)), the approach adopted in this study differs from that adopted in previous studies and has added value by improving the computation of flood hazard of Ninh Kieu district. Furthermore, this study takes a step forward from previous studies, being the first study to probabilistically compute future flood hazard in the study area under climate change. The main value additions of this study, compared to previous studies, are discussed in the following sections.

3.5.1 Difference in using flood probabilities to develop probabilistic fluvial flood hazard maps

One of the biggest differences between the present study and that reported by Apel et al. (2016) is the length of the river discharge time series used for flood frequency analysis. This difference is, in part, due to the different aims of the two studies: Apel et al.'s (2016) aim was to develop flood hazard maps for the present-day while the focus of the present study is to quantify climate change driven variations in the flood hazard.

Consistent with the aim of their study, and following traditional modelling practice, Apel et al. (2016) used flood frequency curves at Kratie of Dung et al. (2015), which were constructed based on the longest possible time series of river discharge at Kratie, spanning 88 years (1924 – 2011). In contrast, however, in studies that focus on climate change driven variations in a given hazard between the present period and future time periods, such as the present study, it is important to ensure that the selected baseline period (and baseline simulations), are in fact representative of the present-day period. This is important because, not only has the climate change signal emerged in several climate variables over the last 50 years or so (i.e. signal is clearly discernible from the inter-annual variability) (King et al., 2015), but also human activities (e.g. reservoirs) have led to noticeable changes in the natural regimes that may have existed earlier in the 20th century (Ranasinghe et al., 2019). Both of these phenomena may change the probability distribution of climate variables over time (Chadwick et al., 2019).

To investigate the stationarity of the upstream river discharge in the Mekong River, the discharge time series at Kratie was analysed, based on the 66 years of data (1924 to 1970 and 2000 to 2018) that has recently been made available at the MRC website. The analysis showed that the peak discharge at Kratie has indeed noticeably decreased over time, and particularly after 2000 (Fig. 3.18, 3.19), likely due to irrigation expansion and upstream dam construction in recent years (MRC, 2010b; Piman et al., 2013).

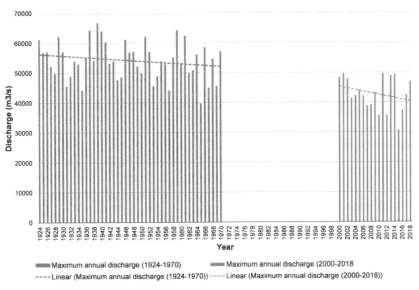

Figure 3.18. Maximum annual discharge at Kratie from 1924 to 1970 and 2000 to 2018

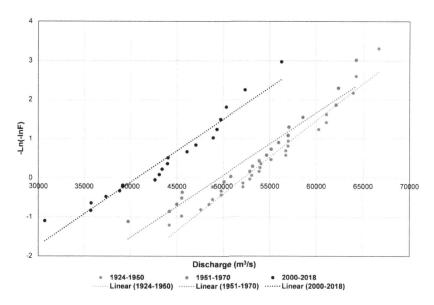

Figure 3.19. Gumbel distribution of discharge peaks at Kratie corresponding to three periods (1924-1950), (1951-1970) and (2000-2018)

The use of the full discharge time series at Kratie to develop flood frequency curves is therefore inappropriate in the present study which aims to quantify climate change driven variations in the flood hazard (Chapter 3) and risk relative (Chapter 4) to present-day conditions, in order to inform the development of climate resilient flood risk reduction measures for the urban centre of Can Tho city (Chapter 5). The use of the full observed discharge data at Kratie, including pre-2000 flow with large flood peaks, can lead to an overestimation of flood hazard and risk. Therefore, for the purposes of this study, only the post - 2000 discharge data were used to represent baseline conditions. This, by necessity, results also in a difference in present-day flood probabilities between this study and that reported by Apel et al. (2016).

Another noteworthy difference between the approaches adopted by the present study as opposed to Apel et al.'s (2016) study arises from the fact that the probabilistic fluvial flood hazard maps for the Ninh Kieu district presented by Apel et al. (2016) were obtained by inputing upstream flood probabilities at Kratie into a combined large-scale inundation model for the entire Mekong Delta developed by Dung et al., (2011) together with a detailed 2D model for the Ninh Kieu district. Flood probabilities at Kratie were then determined based on a bivariate flood frequency analysis using annual extreme discharge and flood volume at Kratie (Dung et al., 2015). However, floods strongly vary over space (Nied et al., 2017; Vorogushyn et al., 2018). This spatial variability of flooding would influence the flood levels at Can Tho which is about 430 km downstream of Kratie. This important aspect is not taken into account by Apel et al.

(2016). Moreover, the river water level at Can Tho and the resulting flood extent and inundation depth in the Ninh Kieu district are affected by the downstream sea level, especially high tides and storm surge (Huong and Pathirana, 2013). Thus, using flood probabilities at Kratie to develop probabilistic fluvial flood hazard maps for the Ninh Kieu district without considering the effect of downstream sea level could lead to large uncertainties in flood hazard computed at Can Tho. The present study overcomes these shortcomings by undertaking 2D flood modelling for Ninh Kieu district based on flood frequency analysis at Can Tho (as opposed to Kratie) and by taking into account both river discharges and downstream sea level in computing river water levels at Can Tho.

3.5.2 The difference in water level and flood extent between two studies

Comparison of the results between the two studies shows negligible differences in estimated present-day water levels corresponding to different RPs between two studies, with the largest difference of 0.05 m corresponding to 100 yr RP. In contrast, the results showed substantial differences in the flood extent corresponding to different RPs for present-day. The inundated area corresponding to 2 yr, 5 yr, 10 yr, 20 yr, 50 yr and 100 yr RP in Apel et al.'s study are 2.37, 3.33, 3.71, 4.30, 4.98, 5.29 km^2, respectively. In contrast, the inundated area for the present-day in this study are 0.42, 0.49, 0.54, 0.60, 0.74, 0.85 km^2, respectively. Apart from the two key methodological differences between the two studies highlighted above, there are also two other reasons that may have led to these differences in estimated present-day flood extents.

While the present study explicitly accounted for the effect of the urban drainage system in Ninh Kieu district on flooding, Apel et al (2016) considered the entire district to be impervious. This has serious implications in terms of flood hazard estimations. The river water level in Can Tho varies following the downstream tidal fluctuation (semi-diurnal tide), as the urban centre of Can Tho (Ninh Kieu district) is connected with the Hau River and Can Tho River via the open sewer channel and urban drainage system. Therefore, for e.g., if during the flood phase of tide, the river water level rises above lowest elevation of the top of the manholes in the city, although without necessarily being higher than the crest elevation of the river embankment, this will lead to flooding in the city centre due to backwater flow through the urban drainage/sewer systems (note: no-return valves are largely dysfunctional in Ninh Kieu district). This is consistent with the flood situation in Can Tho, which was described in Nguyen (2016) (http://www.cantholib.org.vn:84/Ebook.aspx?p=27B9F975353796A6E64627B93B6565 4746C6B65637B91B857557). When the river water level drops during the ebbing phase of the tide, the inundation level is also reduced mostly as flood water is drained through the urban drainage/sewer systems. Hence, incorporating the effects of the flood drainage system, as done in the present study is crucial for correctly estimating flooding in this study area.

Both studies used the DEM presented by Huong and Pathirana (2013) for the study area as the input data of the 2D model. However, stemming from the above mentioned lack of consideration of the effects of the urban drainage/sewage systems, Apel et al. (2016) adjusted the elevation of the DEM data by subtracting 0.5m from the original DEM in order to achieve an acceptable validation their 2D model. Apel at al. (2016) justify this decision referring to the two large fluvial flood events that occurred in 2011, with "extraordinary" peak water levels, but "the banks as given in the DEM were not overtopped, and thus no inundation would occur". However, revisiting the data of water levels at Can Tho station in 2011, used in Chapter 2 to validate the 1D simplified model for the entire Mekong Delta, the peak water levels of these two events occurred on the 28th of September and 27th of October with peak water levels of 2.04m and 2.15m, respectively. Both these water levels are higher than the banks elevation extracted from the original DEM data (approximately 1.9 - 2.0 m) at the surveyed point in Apel et al. (2016). Thus, these two flood events would, in reality, have caused flooding in the Ninh Kieu district by both backflow through the urban drainage system and by direct overtopping of the river embankment. The lowering of the entire DEM is therefore the likely cause for the substantially larger present-day flood extents estimated by Apel et al. (2016), relative to those computed in the present study.

3.6 Conclusions

An efficient modelling approach that combines a simplified 1D hydrodynamic model obtained in Chapter 2 with a detailed 1D/2D coupled model was developed and demonstrated at Can Tho city in the Mekong Delta to produce probabilistic flood hazard estimates. Key features of the modelling approach include (a) Model reduction - a substantially simplified 1D model for the entire Mekong Delta (area of 40,577 km^2) which can simulate one year of riverflow (with an hourly time step) in under 60 seconds (Chapter 2), (b) strategic use of river flood estimates to drive a detailed 1D/2D hydrodynamic flood inundation model (1D/2D model) that is focused on the area of interest (the urban centre of Can Tho city), (c) reduction of the number of 1D/2D coupled model runs required by performing flood frequency analysis while also considering flood hydrograph patterns, and (d) derivation of probabilistic flood hazard quantification for the present and for 2050 under RCP 4.5 and 8.5, with and without land subsidence. The detailed 1D/2D coupled model was successfully validated against measured flood depths at two locations and flood extent at the study area during the Oct 2016 flood event.

Flood hazard maps showing the maximum inundation depth during a flood event were developed for water level return periods ranging from 0.5 yr to 100 yr. Analysis of the flood hazard maps indicates that even under present conditions, more than 12 % of the study area will be inundated by the 100 yr return period of water level. With climate change, but without land subsidence, the 100 yr return period flood extent is projected to more than double by 2050, with not much of difference between the two climate scenarios considered (RCP 4.5 and RCP 8.5). However, if the present rate of land subsidence will continue in the future, by 2050 and under both RCP 4.5 and RCP 8.5, the 0.5 yr and 100 yr return period flood extents are projected to increase by around 15-fold and 8-fold respectively, relative to the present-day flood extent.

These results indicate that reducing the rate of land subsidence, for example, by limiting ground water extraction, would substantially mitigate future flood hazards in the study area. Combining such a measure with a new and more efficient urban drainage network would further reduce the flood hazard. Future modelling studies are needed to quantitatively assess the hazard and risk reduction afforded by these adaptation measures, which could directly feed into risk informed adaptation measures and pathways. For Can Tho, the "do-nothing" management option does not appear to be an option given especially the 15-fold increase in flood extent projected by 2050 for even the twice per year (0.5 yr return period) flood event.

Chapter 4

Assessment of present and future fluvial flood damages and risk for Ninh Kieu district

4.1 Introduction

Assessing flood risk is an essential part of FRM; a critical process for adapting to climate change, economic change and population growth. The main challenge associated with assessing flood risk lies in acknowledging its probabilistic nature, which requires the assessment of flooding at a number of forcing levels, as opposed to assessing the impact due to a pre-determined return period (e.g. 100-year) event, as is commonly done in the engineering-standard based assessments. The main challenge is how to address a large number of external boundary conditions (i.e. river water levels) and corresponding 2D flood hazard simulations to be able to compute the full distribution of inundation depths in a study area. This challenge has been addressed in many studies at the different spatial scales from global to local scale (ICPR, 2001; Evans et al., 2004; Merz and Thieken, 2004; Grunthal et al., 2006; Apel et al., 2009; Adaptation Sub-Committee, 2012; Aerts et al., 2013; Lasage et al., 2014; Ward et al., 2013; Tiggeloven et al., 2020). However, as 2D flood simulations are numerically expensive, such probabilistic risk assessments are computationally demanding. While computational power is not as scarce as used to be, it still hinders the widespread use of risk-based approaches in consultation and decision making contexts. For example, in stakeholder-engaged, co-design contexts where a large number of scenarios are considered and the simulation is an integral part of scenario studies and designs, simulation tools with a very rapid turn-around time (not more than a few minutes) are essential.

Assessment of EADs is a numerically, as well as computationally, intensive modelling exercise that involves the computation of flood hazard (flood depth, velocity, quality, etc.) for a number return periods, which are then used to compute the flood risk and ultimately compute EADs for the current situation and for future scenarios (Rojas et al., 2013; Foudi et al., 2015; Alfieri et al., 2015; Löwe et al., 2018; Adnan et al., 2020). For example, (Löwe et al., 2018) has calculated the flood damages for Elster creek in Melbourne, Australia for various combinations of sea level and rainfall - represented in the form of a mesh - for various adaptation options, which can be used to compute the EAD under various climate scenarios.

This chapter investigates the changes in flood damage and risk (in terms of EAD) due to the different future changes in Ninh Kieu district including the change in upstream river flow, downstream sea level (tide, sea-level rise, storm surge) due to climate change under different future scenarios and land subsidence. Furthermore, this chapter explores the possibility of reusing the flood hazard calculations to represent the flood hazard profile for plausible river water levels (other than those simulated) without performing additional 2D simulations and then computing flood damages in the study area caused by these other plausible river water levels. The purpose of doing this is to generate a wide range of flood hazard and damage maps for the study area for different river water levels with a low computational cost, which can then be used as the underlying database for the co-designing interactive tool presented in Chapter 5, which by design is required to produce rapid visualizations of flood hazard and damage and associated estimate flood damages corresponding to selected input of the user (stakeholders).

This chapter is structured in the following sequence: Section 4.2 presents the research methodology. The results are presented in Section 4.3. It is followed by a discussion in Section 4.4. Sections 4.5 presents the conclusion of this chapter.

4.2 Methodology

Flood risk is defined as the product of the probability of the occurrence of a flood and its consequences (FLOODsite, 2009). Consequences encompasses the harmful effects of a flood, including economic, social, or environmental damage. This study focuses on determining economic losses, including the damages to buildings, furniture and roads.

Calculating EADs

The EAD is shown by the area or the integral under the exceedance probability – damage curve (Grossi et al., 2005; Meyer et al., 2009) (see Fig 4.1).

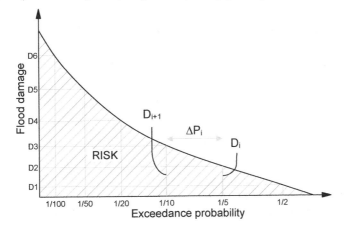

Figure 4.1. Damage-probability curve

A numerical integration method was applied to estimate the EAD for the study area. The EAD is calculated using the formula:

$$EAD = \frac{1}{2}\sum_{i=1}^{n} \Delta P_i (D_i + D_{i+1})$$

(4.1)

71

Where ΔP is the increment of the probability of exceedance, $P = 1/T$, T is the return period under consideration, n is the number of probability levels, D is the damage inflicted.

The methodology adopted in this chapter comprises of two parts: (1) computation of EAD for present and future for Ninh Kieu district using the available flood hazard data obtained in Chapter 3; and (2) reusing the flood hazard from EAD calculations to capture flood hazard profiles for plausible Mekong river levels using the linear interpolation method and then compute flood damages for these events.

To achieve the first goal, inundation depths at all locations in Ninh Kieu district corresponding to the current and future scenarios in probabilistic flood hazard maps obtained in Section 3.4 of Chapter 3 and information on land use of Ninh Kieu for the current and future were combined with stage–damage functions to calculate the damage for the study area. Then total damage corresponding to the different return periods were used to calculate the EAD for each scenario.

To estimate the flood damage due to other plausible water levels at Can Tho, the linear interpolation method was applied to generate flood hazard maps for these flood events based on flood hazard maps corresponding to different return periods (from 0.5 to 100 yr) and a flood hazard map corresponding to the water level of 1.50 m (the water level that is assumed not to cause flooding and damage). Then, the obtained flood hazard maps corresponding to plausible water levels were combined with stage–damage functions to calculate the damage caused by flood events corresponding to these water levels. Linear interpolation has previously been successfully used to approximate the flood prone area based on river stage of Wolf River in Shelby County, the USA without performing additional hydraulic simulations (Javadnejad et al., 2017).

To verify the suitability of the linear interpolation method applied for this study, (i) four flood hazard maps corresponding to 1 yr, 5 yr, 10 yr and 50 yr return period river water levels at Can Tho were used to interpolate and generate flood hazard maps corresponding to 2 yr and 20 yr return period river water level at Can Tho, (ii) the flood hazard maps obtained are compared with flood hazard maps which are obtained directly from the 1D/2D coupled model for 2 yr and 20 yr return period river water level at Can Tho (Fig 4.2). The inundation depth at all locations in the flood hazard map for the 2 yr return period river water level is determined by linearly interpolating between the inundation depth at the corresponded locations of the flood hazard maps for 1 yr and 5 yr return period river water level. Similarly, the inundation depths for the flood hazard associated with the 20 yr return period river water level is interpolated from the flood hazard maps for 10 yr and 50 yr return period river water levels.

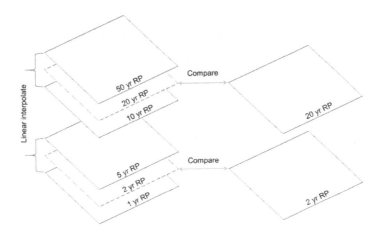

Figure 4.2. Linear interpolation method applied to interpolate flood hazard maps. Rectangles with dashed lines represent flood hazard maps obtained by using the linear interpolation method, rectangles with solid lines represent flood hazard maps obtained directly from the model

4.2.1 Data

Flood hazard maps and scenarios

Probabilistic flood hazard maps with a 15x15 m^2 spatial resolution for Ninh Kieu district corresponding to the current condition and future scenarios in Section 3.4 of Chapter 3 were used to extract the maximum inundation depths at all locations to calculate flood damage and the EAD for the study area. Flood damage and risk corresponding to the current conditions are considered as a baseline to assess the changes in flood damage and risk due to the different future changes. Future scenarios adopted for computing flood damage and risk are shown in Table 4.1.

Table 4.1. Future scenarios adopted for computing flood damage and risk

Scenario	Time	Climate scenario	Land subsidence rate
1	2050	RCP 4.5	1.6 cm/yr
2	2050	RCP 4.5	0
3	2050	RCP 8.5	1.6 cm/yr
4	2050	RCP 8.5	0

Land use maps

Land use maps for 2016 (present condition) and 2050 (under "business as usual" scenario (past is indicative of future trend)) with a resolution of 30x30 m^2 are available for Can Tho city. These maps were developed in the study of Huong and Pathirana (2013), which used Dinamica EGO models to capture the historical rate of land-use transition from 1989 to 2005. These transition rates were combined with known planning-related information to predict land cover maps for Can Tho for the future. In these maps, seven different land-use categories were identified based on satellite images by maximum likelihood classification (supervised), namely, open water, developed low intensity, developed medium intensity, developed high intensity, shrub, grassland and wetland. Land use maps for Can Tho for present and 2050 with seven land-use types are shown in Fig. 4.3.

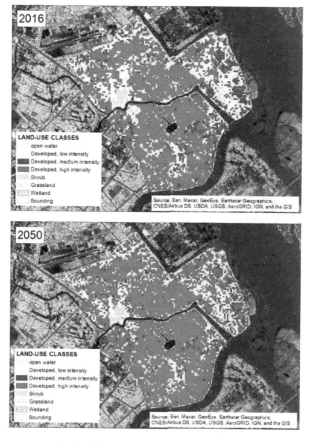

Figure 4.3. Land use maps for Ninh Kieu for present and 2050 (under "business as usual" scenario). Source: Huong and Pathirana, 2013

The projected changes in land-use in Ninh Kieu district in 2050 are relatively small compared to the baseline. This is due to the fact that Ninh Kieu district is located in the central area of Can Tho City with well developed urban infrastructure and high-density urban zones (the red areas) in 2016. The projected changes in land use in the future will be the transition of low and medium density residential areas into the developed high-density residential areas. Also, a small part of the study area is likely to change due to the infill of the green areas covered by grass and shrubs by low and medium density residential areas.

Depth-damage curves

Currently, stage-damage functions for the urban centre of Can Tho city are not available. Therefore, stage-damage functions of Lasage et al. (2014) were used for calculating the EAD for district 4 of Ho Chi Minh City (see Fig. 4.4), were used in this study. Although there are geographical differences between the two study areas (downtown Can Tho and downtown Ho Chi Minh City – both cities are directly administered by the Central Government of Vietnam), they are both located in the south of Vietnam with a distance of about 170 km. However, these two areas have similarities such as located next to a river; both cities are rapidly expanding, were constructed in the same period. Based on these similarities, it is assumed in this study that the typologies of the urban fabric and the values of the assets are very similar. The maximum damage values and damage factors for six typologies as a function of the inundation depth are shown in Fig. 4.4.

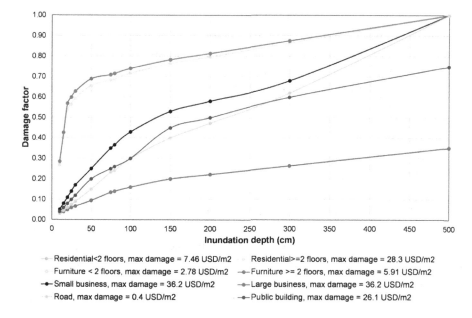

Figure 4.4. Stage-damage curves for six typologies of assets in HCMC based on Lasage et al. (2014). The stage-damage functions for furniture were constructed similarly with the stage-damage functions for two type of houses. It means there is no difference between the shape of the curve of funiture and the corresponding type of house but different about the maximum damage for each type

4.2.2 Flood risk analysis

Calculating flood damage

Damage of a specific asset is calculated using the formula:

$$\text{Damage} = \text{damage function (h)} * D_{max} \qquad (4.2)$$

Where:

D_{max} is max damage value

In order to apply the stage-damage curves of the six typologies of assets (buildings, furniture, road) of HCMC (see Fig. 4.4) to assess the flood damage for the urban centre of Can Tho, the distribution (percentage) of each typology within the seven in each land-use classes of Can Tho (see Fig 4.3) need to be determined. In this study, the damage of the land-use types such as open water, shrub, grassland, and wetland was assumed to be negligible. In this study, the typology distribution of the assets within the land-use classes was assessed using satelite imagery of Google Maps. Five areas of each land-use class (low, medium and high density) have been selected randomly from the land use map of the urban centre of Can Tho city. The distribution of each class of assets (i.e. area of each typology of asset) within the above locations was determined. Average values of the five selected areas were used.

The damage per square metre or each grid cell corresponding to each land-use class of Can Tho was calculated by summing up the damage of each typology of asset (roads, buildings, and furniture). The damage of each assets typology was calculated by multiplying the max damage of the corresponding asset with the corresponding damage factor related to the inundation depth at that location. For cells on the flood inundation maps with a value of the maximum inundation depths below 0.02 m are indicated as no inundation (no risk).

The total damage for the study area corresponding to each return period of water level (0.5, 1, 2, 5, 10, 20, 50, 100 yr) was calculated by aggregating the damage at all grid cells. Due to the difference in spatial resolution between the hazard map (15x15 m^2) and the land use map (30x30 m^2), the spatial resolution of flood hazard maps was resampled from 15x15 m^2 to 30x30 m^2.

Calculating EADs for Ninh Kieu district

Total damages corresponding to different return periods (0.5, 1, 2, 5, 10, 20, 50, 100 yr) of each scenario obatained above were used to calculate the EAD for the baseline and future scenarios for Ninh Kieu district.

4.3 Results

This section presents the flood damage maps and the EAD for the urban centre of Can Tho city corresponding to the baseline (current situation) and the future scenarios under RCP 4.5 and RCP 8.5 (without and with land subsidence). Flood hazard and damage maps, as well as estimated flood damage for other plausible flood events that are calculated using linear interpolation, are also presented in this section.

4.3.1 Flood damage maps and the EADs corresponding to the baseline and future scenarios

Figure 4.5 shows the distribution of the damage of the case study area for flood events corresponding to 5 and 100 yr return period of water levels at Can Tho for the baseline, and 2050 under RCP 4.5 and RCP 8.5 (without land subsidence) (scenarios #2 and #4 respectively). The remaining flood damage maps which correspond to 0.5, 1, 2, 10, 20 and 50 yr return period water levels for these scenarios are shown in Figs B1 – B3 of Appendix B.

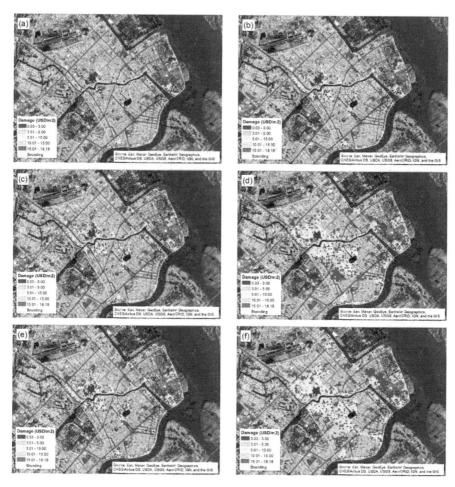

Figure 4.5. Flood damage maps for flood events corresponding to 5 yr and 100 yr return period of water level, Figs a and b for the baseline, Figs c and d for scenario #2, Figs e and f for scenario #4

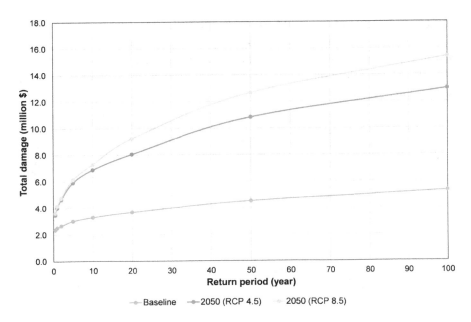

Figure 4.6. Flood damage curves for the baseline, and for future (2050) under RCP 4.5 and RCP 8.5 (without land subsidence), respectively

Figure 4.5 highlights some locations with high flood damage in the study area for all scenarios. These locations correspond to the locations with high inundation depths on the flood inundation maps presented in Chapter 3.

For the baseline, the total damage caused by a flood event corresponding to a 100 yr return period of water level, is more than double the flood damage corresponding to 0.5 yr return period of water level: the total damage increases from 2.4 million USD (0.5 yr return period) to 5.3 million USD (100 yr return period) (see Fig 4.6 and Table B1 in Appendix B). The EAD for the baseline is 5.3 million USD/yr, which is equal the total damage, caused by a flood event corresponding to a 100 yr return period water level at Can Tho.

In comparison, for the future scenarios (without land subsidence), the total damage for a flood event corresponding to a 100 yr return period of water levels at Can tho under RCP 4.5 and RCP 8.5 is 13.0 million USD and 15.4 million USD, respectively. The EAD for these scenarios (under 4.5 and RCP 8.5) are 8.9 million USD/yr and 9.3 million USD/yr, which is 1.7-fold and 1.8-fold higher than the EAD for the baseline, respectively.

Figures 4.7 shows the distribution of damage of the case study area for flood events corresponding to 5, 100 yr return period water levels at Can Tho for the baseline, and for 2050 under RCP 4.5 and RCP 8.5 (with land subsidence) (scenarios #1 and #3 respectively). The remaining flood damage maps which correspond to 0.5, 1, 2, 10, 20 and 50 yr return period water levels for these scenarios are shown in Figs B4, B5 in Appendix B. When land subsidence is taken into account, the flood damage in the study areas increases markedly.

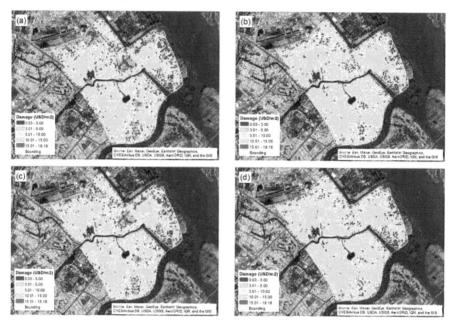

Figure 4.7. Flood damage maps for flood events corresponding to 5 yr and 100 yr return period of water level, (a and b) for scenario #1, (c and d) for scenario #3

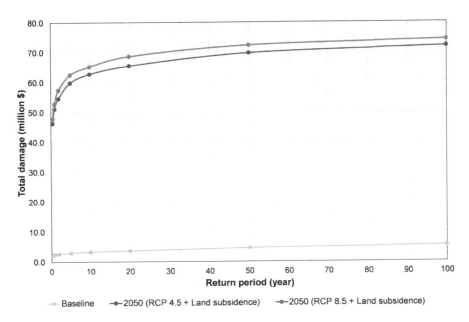

Figure 4.8 Flood damage curves for the baseline, and for future (2050) under RCP 4.5 and RCP 8.5 (with land subsidence), respectively

Figure 4.7 indicates that the majority of the study area will be severely damaged by flood events corresponding to all return periods of water level at Can Tho in the future scenarios under RCP 4.5 and RCP 8.5 (with land subsidence). The total damage caused by a flood event corresponding to a 100 yr return period of water level under both RCP 4.5 and RCP 8.5 is 72.0 million USD and 74.2 million USD respectively (see Fig 4.8 and Table B2 in Appendix B), which is approximate 6 times and 5 times higher than that without land subsidence, respectively. The EAD for the future scenario under RCP 4.5 and RCP 8.5 (without and with land subsidence) varies from approximate 2-fold to 21-fold compared to the EAD for the baseline, respectively.

These results can be used to assess how several large scale adaptation measures could affect the EAD. Here 3 such adaptation measures are considered:

(1): control land subsidence and greenhouse gas emission is kept at high level (RCP 8.5).

(2): control land subsidence but greenhouse gas emission is reduced to a lower level (RCP 4.5).

(3): No control land subsidence and greenhouse gas emission is reduced to a lower level (RCP 4.5).

Table 4.2 shows that if land subsidence is controlled by for example regulations to restrict groundwater extraction, the EAD could potentially be decreased by 91.4 % in 2050. If greenhouse gas emissions will be reduced, such that follows RCP 4.5 rather RCP 8.5, then a further 95.4 % reduction of the EAD could be achieved by 2050. If only the mitigation target of RCP 4.5 is met without changes in subsidence rate, the EAD will decrease by 3.8 %.

Table 4.2. Expected annual damage for 2050 (under RCP 4.5 and RCP 8.5) (without and with land subsidence) and percentage reduction of the EAD corresponding to each adaptation and miltigation measure

Scenario	EAD (Mil. USD/year)		% Reduction of EAD when applying adaptation + miltigation measures		
	Without LS	With LS	(1)	(2)	(3)
2050 (RCP 4.5)	8.9	104.4	91.4	95.4	3.8
2050 (RCP 8.5)	9.3	108.5			

4.3.2 Flood hazard and damage maps by linear interpolation

Figures 4.9 and 4.10 present flood hazard and damage maps for the baseline corresponding to 2 and 20 yr return period of water level at Can Tho obtained directly from the coupled 1D/2D model and by linear interpolation, respectively.

Figure 4.9. Flood hazard maps for the current situation (baseline), Figs a and b correspond to the 2 yr return period of water level directly from the model (Fig a) and linear interpolation (Fig b), Figs c and d - the same comparison corresponding to the 20 yr return period of water level. The inundated area corresponding to the 2 yr return period of water level is 0.42 km^2 as obtained directly from the model, and 0.48 km^2 by linear interpolation, respectively. While for the 20 yr return period, the inundated area for each method is 0.60 km^2 (direct model result) and 0.74 km^2 (linear interpolation)

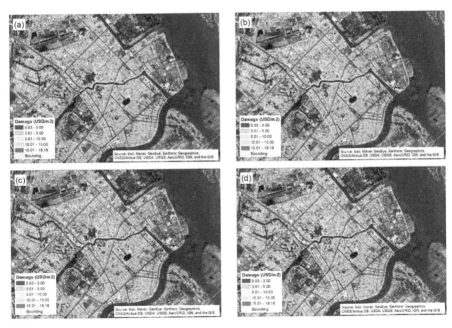

Figure 4.10. Flood damage maps for the current situation (baseline), Figs a and b correspond to the 2 yr return period of water level directly from the model (Fig a) and linear interpolation (Fig b), Figs c and d - the same comparison corresponding to the 20 yr return period of water level. Total flood damage corresponding to the 2 yr return period of water level is 2.67 million USD as obtained directly from the model, and 2.7 million USD by linear interpolation,

*respectively. While for the 20 yr return period, total flood damage for each method is 3.7
million USD (direct model result) and 3.8 million USD (linear interpolation)*

The inundated area based on the linear interpolation method is 0.06 km² and 0.14 km²
larger than the inundated area obtained directly from the model corresponding to the 2
and 20 yr return period (a difference of 14.3 % and 23.3 %), respectively. However, the
difference in the inundation depth at the extended areas between two methods is
insignificant for both return periods: it concerns only few centimetres. This is reflected
by the insignificant difference in the total flood damage between two methods: 2.67
million USD and 2.7 million USD corresponding to the 2 yr return period, and, 3.7
million USD and 3.8 million USD for the interpolation method corresponding to the 20
yr return period (a difference of 1.1 % and 1.6 % for the 2 yr return period and 20 yr
return period, respectively). The differences in the flood extent and inundation depth, as
well as in the total flood damage, tend to become smaller when the interpolation method
is used to calculate for flood events corresponding to lower water levels. Hence, the
linear interpolation method appears to be an acceptable solution to generate flood
hazard maps and calculate flood damages for flood events corresponding to plausible
water levels.

4.4 Discussion

Table 4.3 summarises some main metrics corresponding to a 100 yr RP flood event and
EAD for the baseline period and for 2050 (under RCP 4.5 and RCP 8.5.

*Table 4.3. Summary of the inundated area, maximum inundation depth, and total damage
corresponding to a 100 yr return period flood event and EAD for present-day and 2050 (under
RCP 4.5 and RCP 8.5) (with and without land subsidence)*

Indicator	Present	2050 (RCP 4.5)		2050 (RCP 8.5)	
		Without LS	With LS	Without LS	With LS
Inundated area (Km²)	12.7	29.2	95.2	34.2	96.4
Maximum inundation depth (m)	2.52	2.63	3.17	2.66	3.20
Total Damage (Mil. USD)	5.3	13.0	72.0	15.4	74.2
EAD (Mil. USD/year)	5.3	8.9	104.4	9.3	108.5

The results in Table 4.3 show that the effect of land subsidence is significantly larger
than the effect of climate change on inundation depth and flood extent at Ninh Kieu
district. Specifically, in 2050, 0.11m and 0.14m are the increased inundation depths

caused by a 100 year RP flood event compared to the present due to the effect of climate change under scenarios RCP 4.5 and RCP 8.5, respectively. While this figure increases by 0.54m to 0.65 and 0.68 when further considering the effect of land subsidence. As a result, land subsidence contributes 80% and 74% of the additional inundated area compared to the 20% and 26 % due to the impact of climate change corresponding to scenario RCP 4.5 and RCP 8.5, respectively.

From the EAD computations presented above, it is evident that there is considerable flood risk at present and in the future in Ninh Kieu district. The effect of climate change on river flow and sea level is projected to lead to a substantial increase in the flood risk at Ninh Kieu (1.7-1.8 times) in 2050 compared to present-day. This is consistent with the study of Lasage et al. (2014) who reported an increase of 112-115 % in the EAD in district 4, Ho Chi Minh City due to only sea-level rise, and de Moel et al. (2014) who projected an increase of 1.5 times in the EAD by 2050 due to the combined effect of climate change driven variations in peak river discharge and sea-level rise) for the case study of Rotterdam in the Netherlands. The flood risk in Ninh Kieu district is projected to increase (20-21 times) when the combined effects of climate change and land subsidence are taken into account. This is similar to projections of increasing flood risk in some cities around the world (e.g. Shanghai, Tokyo, Bangkok and Jakarta) which are confronted with land subsidence (Kaneko and Toyota, 2011; Ward et al., 2011b; Erkens et al., 2015).

The re-use of flood hazard maps underlying the EAD assessment to create a wide range of hazard and damage maps for other plausible river levels on the Mekong River using linear interpolation significantly reduces the computational cost, as this precludes the need to perform additional 2D simulations. Flood hazard and damage maps obtained are used to create a rich database for an interactive tool (in Chapter 5, which can be used in co-design meetings to produce rapid visualizations (flood hazard and damage maps) and to estimate flood damages corresponding to selected inputs of the user (stakeholders), which would help them understand event-based hazards at specific locations of interest.

However, the use of stage-damage functions that were developed for Ho Chi Minh City to calculate the flood damage and risk for Ninh Kieu district may lead to discrepancies between actual and predicted damage at Ninh Kieu corresponding to a specific flood event. This is because the asset value and the level of damage associated with each inundation depth of each typology of assets are different between the two locations. Additionally, the use of satellite imagery (Google Maps) to determine the typology distribution of the assets within the land-use classes in Ninh Kieu district is not highly accurate. This also may lead to a difference between actual and predicted damage at Ninh Kieu district. However, as stage-damage functions for different typologies of assets are currently not available for Ninh Kieu district, the results presented here constitute a best-effort estimation.

4.5 Conclusions

Flood damage maps and EADs for the urban centre of Can Tho corresponding to present and future scenarios (which take into account the effect of climate change and land subsidence) are developed and computed based on probabilistic flood hazard maps (produced in Chapter 3) for the study area and the stage-damage curves of Lasage et al. (2014) used for calculating the EAD for district 4 of Ho Chi Minh City. The current flood risk of the urban centre of Can Tho City, expressed in EAD is 5.3 million USD/year. The flood damage in the study area is projected to increase in the future due to the impact of climate change on upstream river flow, downstream sea level (tide, sea-level rise, storm surge). The expected damage in 2050 under two climate scenarios RCP 4.5 and RCP 8.5 will increase by 1.7 times and 1.8 times compared to the baseline (2016). Land subsidence has a more significant effect on the EAD than the impact of climate change, the expected damage in the study area in 2050 under RCP 4.5 and RCP 8.5 when land subsidence is accounted for in the calculations is approximately 20-fold and 21-fold higher than the baseline.

This chapter explored the possibility of reusing the flood hazard calculations of EAD assessment to capture flood hazard profiles for plausible river water levels without performing additional expensive 2D simulations by using the linear interpolation method and then compute flood damages in the urban area caused by these events. The flood hazard and damage maps underlying the EAD calculations were re-used to generate flood hazard and flood damage maps for two other flood levels corresponding to the 2 yr and 20 yr return period river water level to verify the suitability of the linear interpolation method applied for this study. The results obtained show that: (i) the linear interpolation method is acceptable for generating flood hazard and damage maps and for computing flood damages associated with flood events corresponding to other plausible river water levels with a low computational cost, (ii) the accuracy of the linear interpolation method is higher for flood events corresponding to lower river water levels.

With the ability to generate flood hazard and flood damage maps for plausible water levels at Can Tho from the current EAD dataset increase the possibility to create a wide variety of scenario-based narratives, such as the impact of climate change forcing (river flow, sea-level rise, storm surge); land subsidence; and, the effectiveness of adaptation measures. Therefore, the EAD computations performed here can be reused to assess flood hazards and damages for exploratory planning of various FRM strategies by urban planners and decision-makers.

Chapter 5

Developing and testing an interactive, web-based flood risk management tool for co-design with stakeholders[3]

---

[3] This chapter is partially based on Ngo, H., Radhakrishnan, M., Ranasinghe, R., Pathirana, A., Zevenbergen, C.: Instant Flood Risk Modelling (*Inform*) Tool for Co-design of Flood Risk Management Strategies with Stakeholders, Water, submitted, 2021.

5.1 Introduction

About 250 million people in the world are affected by floods every year (UNISDR, 2013), and the annual average economic losses have exceeded 40 billion USD in recent years (OECD, 2016). Developing flood risk reduction strategies to minimise damage caused by floods is essential in FRM. Collaborative learning and designing among stakeholders is an increasing trend in local FRM (van den Belt, 2004; Beall and Zeoli, 2008; Suarez et al., 2009; Den Haan et al., 2020). For the selection and implementation of flood risk reduction measures, the importance of stakeholder engagement (e.g. citizens and interest groups, businesses, officials, and decision-makers, etc.) is widely acknowledged. Stakeholders are also increasingly involved in the design phase of measures supported by user-friendly flood risk models, such as in the Dutch *Room for the River* program (Rijke, 2014), and SimDelta (Rijcken et al., 2012; Rijcken, 2017). The effectiveness and success of co-designing sessions have been demonstrated by the *Blokkendoos* tool (WL Delft hydraulics, 2003; Zhou et al., 2009) used within the *Room for the River* program in the Netherlands. In interactive sessions supported by the *Blokkendoos* tool, stakeholders participated in the design process using 'what-if scenarios' to explore the impact of various interventions on the flood level as well as the flood risk in the study area, which allowed them to prioritize interventions and to engage in an inclusive decisionmaking process. Effective participation of stakeholders in interactive work sessions requires a fast and accurate modelling system with a user-friendly interface, which can simulate the user's interventions to provide outputs that can be understood by all stakeholders, especially those who are non-water specialists (Den Haan et al., 2020). These features have been identified as the most sought after attributes of flood simulation models to support practitioners in flood disaster management (Leskens et al., 2014).

This chapter presents an interactive, web-based tool, *Inform*, which is developed based on a simplified 1D model for the entire Mekong Delta (Chapter 2), flood hazard and damage maps, as well as on estimated impacts (direct damage) (Chapter 3 and Chapter 4). *Inform* is designed with the aim of satisfying the above mentioned user requirements, allowing users to assess different scenarios and receive visual outputs within one minute. Pilot testing with experts from universities and institutes in the Netherlands and Vietnam and colleagues (PhD fellows) at IHE Delft to evaluate the tool's key features (e.g. user interface, the effectiveness of the tool and its outputs) was conducted. For the evaluation, seven criteria were used, which were extracted from a literature review. *Inform* is used here to demonstrate how it maybe used to support rapid flood risk assessment to facilitate a co-designing approach aiming at exploring, identifying and selecting flood risk reduction measures for the urban centre of Can Tho city (Ninh Kieu district) with the stakeholders' participation.

The content of this chapter is structured as follows: Section 5.2 presents the results of a literature review on the challenges of generating interactive tools to support FRM. Addressing challenges using an interactive tool for Can Tho is presented in Section 5.3. Pilot testing and evaluation of *Inform* is presented in Section 5.4, followed by a discussion in Section 5.5. Section 5.6 presents conclusions of this chapter.

5.2 Challenges in creating interactive tools to support FRM

There are freely available tools in countries such as The Netherlands and UK, that give information on long term flood risk for an area, possible causes of flooding and how to manage flood risk (Rijkswaterstaat, 2020; UK Environment Agency, 2020). Also, organisations such as Melbourne Water in Australia issue flood level certificates that contain information about flood level and the probability of occurrence of flooding at the property level (Melbourne Water, 2019). These are awareness creation or risk information dissemination platforms that are meant to present expert domain knowledge in a form that the general public can easily understand. The effectiveness of FRM measures is an important strand of information that can be visualized and which sits at the interface between science, policy and implementation. The visualization can be in the form of a simple infographic or a detailed three-dimensional real-life virtual tour of the FRM measures under consideration.

The common feature among all of the aforementioned online platforms and interactive visualisations is that the entire process of hydraulic modelling is completed beforehand – updated periodically – with or without the possible combination of FRM measures and is stored in a database. Some online tools have an interactive front-end that allows the user to select a postcode for which the flood risk and probability of flooding are retrieved from the database and results are displayed instantly either in the form of a map or as text (e.g. Rijkswaterstaat (2020); UK Environment Agency (2020)). However, these tools do not have options for the user to perform 'what-if' scenarios to explore the impact of various interventions on the flooding level as well as the flood risk in the study area. Ideally, a platform should allow the user to change inputs such as river flow rate, rainfall intensity, water level and FRM measures through an easy-to-use graphical user interface (GUI); run a fast hydraulic model to determine the flood extents, duration and depths; use this information for calculation of immediate risks such as flood damage and long term risks; and display the information in the form of online maps, tables and graphics. These outputs should be understandable to stakeholders, especially those who are non-water specialists. It must also be ensured that the tool is fast enough to perform simulations of a variety of scenarios based on selected user inputs to assess the flood risk reduction measures during a co-design work session. These are must-have features of co-design tools identified from a number of stakeholder interviews (Leskens et al., 2014).

The data preparation, processing and prerequisites have to be taken into account to develop an online flood risk tool that can help generate the details that are needed to create different flood narratives for various stakeholders during co-design work sessions.

Following from the above, it is clear that tool developers need to comply with a set of minimum requirements in order to create a user-friendly flood risk tool. These requirements are discussed below.

Ensuring reliability of tool outputs: The reliability of the tool is of utmost importance to the user (Leskens et al., 2014). Confidence in the outputs of modelling tools is paramount for the uptake of the tool by various stakeholders. For example, the user must have confidence that simulated flood depths and flood damages in the tidal urban sub-catchment of the river are reliable.

Ease of use and avoiding input overkill: Differences often exist in expertise and interest of stakeholders' participating in co-design work sessions (Faulkner et al., 2007; Timmerman et al., 2010; Samuels, 2012; Leskens et al., 2017). These require that tool developers must have an understanding of the knowledge level and information needs of potential users in order to use the tool effectively. Questions such *as "does the tool require the user to bring in data?"* and *"should the user select all of the parameters to run the tool?"* have to be considered by the developers at the tool design stage. Lack of detailed knowledge of the (local) hydrological context and technical aspects (for instance about the specifications of interventions), difficulties in understanding the instructions and cognitive loading should not provide an impediment for a wide range of stakeholders to use the tool (Scheffran, 2007; Aubert et al., 2018).

Time taken to generate tool outputs: Co-design work sessions typically last a few hours (Mintzberg et al., 1976), therefore, they call for models to be fast enough to provide multiple outcomes and foster iteration during a single work session (Leskens et al., 2014). Questions such as *"how long does the user need to spend on selecting the input parameters and how long does the user have to wait to get the outcomes?"* and *"what is the acceptable waiting time between the selection of inputs and the display of results?"* have to be considered in the development phase of the tool.

Transcending coarser and finer resolutions across spatial scales: A flood risk tool that allows the user to identify the flood risk and to assess the effectiveness of adaptation actions across various locations and spatial scales will contribute to the flood risk knowledge base of the various stakeholders active in the delta or catchment area. This in turn offers conditions for collaboration and joint actions. The following questions seem to be relevant in this context: Is the tool capable to simulate flooding across spatial scales from a catchment of the size of the Mekong Delta and to an urban area whereby the information (flood delineation and damages) provided has sufficient resolution to support multi-level governance decision-making? For example, the knowledge of local people (including their perceived vulnerability) on the effectiveness and feasibility of adaptation actions they can take individually at the local scale or collaboratively at a larger spatial scale may incentivise them to act and thus increases their ability to frame, understand and influence their flood and climate change risks (Maloney et al., 2011; Ngo et al., 2020a).

Transcending coarser and finer resolutions across temporal scales: What are the time scales relevant for the tool? A tool which supports high-level strategic decision making and is able to explore pathways to future flood protection, warrants to simulate flooding and to compute flood damages covering a time horizon of at least 50 years given the long lead times of large interventions. In addition, stakeholders' engagement and responses are influenced by how risk is presented and the factor time is very important in risk communication as there is a physiological distance or disconnect of long term risks or impacts such as climate risk (Brügger et al., 2016; Spence et al., 2012). Hence it is imperative to generate data to elucidate the long term potential impacts and the lead time of adaptation actions.

Interpretation and relevance of tool outputs across a wide spectrum of stakeholders: The tool outputs should support decision making at the strategic policy planning level, project planning and implementation level, and also at the operational level to create awareness among the general public? It is of utmost importance that the stakeholder needs are identified and what they actually want to know. Moreover, the stakeholders should have the ability to interpret the outputs (Leelawat et al., 2013)? The latter often requires customized guidance.

Assessing the effectiveness of FRM measures: Delivery of risk information without suitable actions to minimise or eliminate risk can cause concern and anxiety among stakeholders (Leiserowitz, 2006; Nisbet, 2009). How effective are the planned FRM measures in reducing flood damages? Can the effectiveness of FRM initiatives at multiple levels be assessed using the tool? For example, the effectiveness of FRM actions from the Mekong Delta Plan (MDP, 2013) is crucial information at the strategic planning level, the effectiveness of retrofitting and elevating the floor level of houses as property level is relevant information for the City Committee and citizens, respectively.

Addressing the aforementioned challenges well provide the conditions required for the development and acceptance of a useful flood risk gaming tool to support FRM. Some of the challenges, such as reducing the waiting time can be resolved using computational techniques. In contrast, challenges such as integration of FRM measures need to be dealt from the perspective of FRM planners, and challenges such as versatility need to be understood from a human-machine interface perspective.

5.3 A new co-designing interactive tool to address flooding related challenges in Can Tho

The instant Flood risk modelling tool (*Inform*) for Can Tho developed here is an interactive web-based tool which provides an estimate of flood water level at Can Tho, generates flood inundation (flood extent and inundation depth) and flood damage maps, as well as estimated flood damage for Ninh Kieu district in Can Tho city in a short time corresponding to a single flood event based on user-defined inputs (https://fg.srv.pathirana.net/) (Ngo et al., 2020b). The underlying flood hazard and risk data are extracted from Chapter 3 and Chapter 4 while users are provided with the option of testing how certain adaptation measures may affect the flood hazard and/or risk. *Inform* covers the entire Mekong Delta, including Can Tho city where has an important demographic and economic significance (DWF, 2011; SCE, 2020). The background engine of *Inform* comprises a simplified, calibrated and validated 1D SWMM model for the entire Mekong Delta to calculate the water levels of the Mekong River at Can Tho (Ngo et al., 2018); an output library comprising grid-level inundation depths directly from 1D/2D coupled PCSWMM simulations and via the interpolation method for Ninh Kieu district of Can Tho city for a wide range of Mekong River water levels at Can Tho; a grid level flood damage calculation model based on depth-damage curves for the different type of assets in Ninh Kieu district. *Inform* uses the modelled water levels at Can Tho to generate the flood hazard and compute the associated flood damages. Reliability of the 1D/2D coupled flood model results has been established by calibrating the model outputs against measured inundation depths and flood extent at the study area during historical flood events in Section 3.4 of Chapter 3. Screenshots of inputs and outputs of the *Inform* are presented in Fig 5.1.

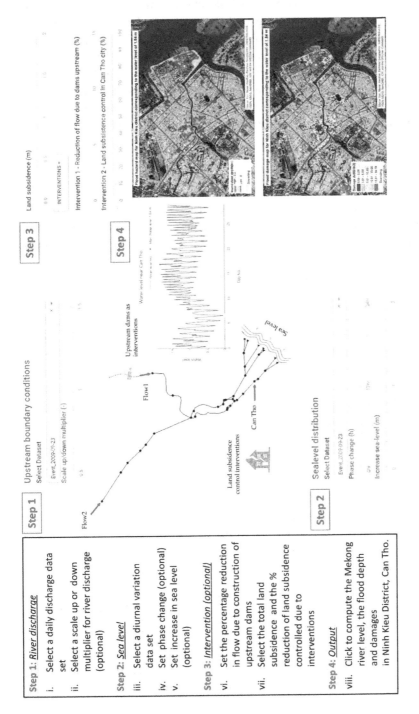

Figure 5.1. Flood risk tool (Inform) for Can Tho - inputs and

Inform outputs can be used to create authentic quantitative information (e.g. flood water level, flood hazard and damage maps, as well as estimated food damage corresponding to each flood event) to inform, explore and strategize with a number of stakeholders through the process of collaboration, critical thinking and creativity. This can lead to outcomes, such as awareness, validation, trust, policies, innovations and futuristic planning, etc. How *Inform* addresses the user-identified challenges in creating the interactive tool and how it could be used in FRM domains such as strategic planning, city administration, project implementation and community engagement in Can Tho are explained below.

Ease of use and avoiding input overkill

The necessary input data for *Inform* is already processed and stored in a database. Hence the user does not need to collect additional data, which avoids input overkill. From the *Inform* database, the user selects only six input parameters through a graphical user interface to generate the outputs. Two additional selections have to be made in case if outputs for FRM intervention are required. The user inputs to *Inform*, which can be selected from the existing database using a drop-down menu, and data ranges are:

(i) upstream boundary condition of Mekong River at Kratie - i.e. flows at upstream of TonleSap (left side) and river flows at Kratie in the Mekong River (right side) and a scale factor to change the river flows. The scale factor is the ratio that the user wants to change the original river flow by, and here it is limited from 0.5 to 1.5 times compared to the original flow. The scale factor is used when the user wants to estimate the change in the river flow due to the effect of climate change to compute the resulting flood water level in Can Tho and the associated flood damage in Ninh Kieu district in the future compared with the past flood event.

(ii) sea level at the Mekong river mouths, which are the downstream boundary condition at East Sea – i.e. the diurnal variation of sea level for select dates from the drop-down menu, phase change of diurnal variation (the phase change represents a future change in the shape of sea level time series during a flood event, which can lead to changes in the river water level in Can Tho and the resulting flood damage in Ninh Kieu district) and increase in the amplitude of sea level (the change in amplitude represents a future increase in sea-level rise due to the effect of climate change). The change in the amplitude of sea level is limited from 0 to 2 m, and the users only can select a value in this range.

(iii) land subsidence at Can Tho, which reduces the elevation of the ground surface. As a result, the inundation depth in Ninh Kieu district will increase when computing flooding, resulting in an increase in the flood damage.

(iv) FRM interventions – upstream interventions such as dams and reservoirs that can reduce a certain percentage of river flow (input), and avoiding damages by controlling land subsidence, where the rate of land subsidence and percentage reduction in land subsidence are inputs. Upon selection of these inputs, the user can activate the hydraulic models and flood damage calculation models with only one click of the mouse.

Time taken to generate tool outputs

The time taken to make choices of the eight input parameters (six basic parameters and two for FRM interventions) required for *Inform* is fast (a few minutes), for a user conversant with standard inputs to a hydraulic model. After these inputs are selected, it takes about one minute for *Inform* to generate the following outputs: (i) the maximum water level in the river and its timing at Can Tho; (ii) map showing the maximum flood depth in every grid cell in Ninh Kieu district; (iii) the flood damage due to inundation in Can Tho in million USD; (iv) map showing the damage in every grid cell due to inundation in Ninh Kieu district in USD per square meter.

Inform can theoretically generate the maximum water level in the river for any combination of the river flow and sea level. However, at present, for combinations that provide a water level at Can Tho higher than 2.72 m, *Inform* can only generate flood hazard and damage maps and total damage corresponding to the water level of 2.72 m. This is because the current database of Inform only contains hazard and damage maps, and total damages corresponding to the 100 yr return period flood event under scenario RCP 8.5 with the highest water level of 2.18 m combined with a land subsidence value of 0.54m at Can Tho in 2050..

Transcending coarser and finer resolutions across spatial scales

The *Inform* output can help the strategic planners to determine optimal protection levels for FRM infrastructure, preparedness measures for the study area and develop appropriate policies to support FRM and strategize delta planning (Seijger et al., 2019). More specifically, the river water level generates maximum inundation depths at grid level in the city centre (Fig 5.2) and will help identify the flood hazard in those grid cells, which would help the determination of flooding hotspots. The flood hazard map can also help the city officials to identify critical infrastructure and community at the risk of flood by overlaying the inundation map on their critical infrastructure and demography GIS database. Grid level flood hazard details from *Inform* can lead to the development of flood hazard ranking of grid cells or community pockets, which is a vital piece of information to channelize relief actions in the event of flooding. This will aid the city administration to contextualise the flood risk problem in Can Tho and help planning, implementation and coordination of FRM and relief measures in the near future or in the long term (Radhakrishnan et al., 2017).

Flood inundation

Estimated damage due to the flood event 2.73 mil. USD

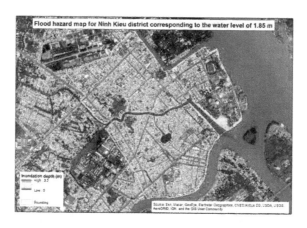

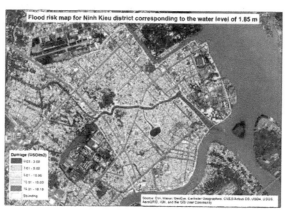

Figure 5.2. The inundation map, flood damage map, as well as the estimated total flood damage due to the flood event for Ninh Kieu district, Can Tho city

Inform calculates the total flood damage in Ninh Kieu district for the selected river flow – sea level combination, and the map illustrates the grid level flood damages in USD per m² (Fig 5.2). From this map, it can be inferred that the damages vary across the grids in Ninh Kieu district. At the coarser scale, the total flood damage information can be useful for strategic delta planning as it can help policy makers understand the impact of flooding in monetary terms in the most populous city and the hub of economic activity in the Mekong Delta. The total damages for different river flow – sea level combinations can help the city administrators to establish the range of monetary loss due to the past events and can help them prepare in terms of creating a reserve to compensate for future losses or to start formulating plans for FRM interventions to avoid or minimise flood risk.

Transcending coarser and finer resolutions across temporal scales

The river flow data and sea level data in *Inform* tool contains actual data from past events, which can generate historical flood events. However, what makes *Inform* interesting is the scale factor that can be applied to the river flows and increase in sea level that enables the user to make the changes to the historic data to explore 'what if scenarios' of present and future. The effect of climate change through the increase in sea level at the Mekong estuary mouths and the river flows was clearly evident through different future flood extents in Can Tho for climate scenario such as the IPCC climate scenarios (i.e. RCPs). As mentioned earlier, the data for the *Inform* output library was obtained through the 2D modelling of flood depths and extent for various levels of Mekong River level. The timing of occurrence of the plausible future water levels of Mekong River in various IPCC scenarios was obtained from discharge projections calculated by Hoang et al. (2016). For example, with the help of *Inform*, the user can adjust the inputs to compute what would be the estimated flood damage of an event that has a similar return period of the flood event in the year 2000 flood if it occurs in 2050 under RCP4.5 or RCP8.5. The flood in the year 2001 was a 1 in a 0.3-year event. In the future, due to the impact of climate change, a 1 in the 0.3-year event would correspond to an increase in 20 % of upstream flows (compared to the year 2001 flood flow) together with a 0.2 m and 0.3 m increase in downstream sea level, in the RCP 4.5 and RCP 8.5, respectively. These change in river flows and sea levels can be given as inputs in *Inform* by using the scale-up/down and increase in sea level options, and the corresponding river level and flood damages can be obtained (Table 5.1).

Table 5.1. The maximum water level in Can Tho and estimated damage in Ninh Kieu district for historical flood events in 2001 (1 in 0.3 years) and flood events of same return period reoccurring in 2050 based on user-defined inputs under RCP 4.5 climate scenarios and RCP 8.5

Selected event	Past		2050 (RCP 4.5)		2050 (RCP 8.5)	
	Maximum water level (m)	Estimated total damage (Mil. USD)	Maximum water level (m)	Estimated total damage (Mil. USD)	Maximum water level (m)	Estimated total damage (Mil. USD)
The 2001 flood event (1 in 0.3 years)	1.75	2.53	1.99	4.70	2.07	7.72

Users with access to the river flow data, such as city council, irrigation department and researchers can use the tool to create comparative information based on the Mekong River water level for various return periods. For example, the flood extent in the year 2020 in Ninh Kieu district for a 1 in 100 year return period Mekong water level (2.15 m) is 12.7 % (Fig 5.3). However, the flood extent in the year 2050 in Ninh Kieu district for a 1 in 100 year return period Mekong River water level were found to be 29.2 % and 34.2 %, under RCP 4.5 (Fig 5.4) and RCP 8.5 (Fig 5.5) respectively.

Figure 5.3. Inundation map corresponding to 100 year return period water level at Can Tho in the year 2020

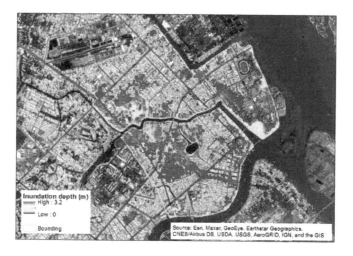

Figure 5.4. Inundation map corresponding to 100 year return period water level at Can Tho in the year 2050 under RCP 4.5

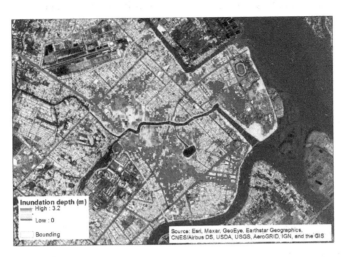

Figure 5.5. Inundation map corresponding to 100 year return period water level at Can Tho in the year 2050 under RCP 8.5

Thus, users from the planning and program formulation domain can easily relate the inputs values to future conditions using the available information on climate change in the Mekong Delta and create useful information from the outputs. This would also enable the policymakers to explore strategic planning and policy narratives, such as adaption pathways for Can Tho (e.g. Radhakrishnan et al. (2018b)) to adapt to changing climate and mitigate these risks.

Interpretation and relevance of outputs across a wide spectrum of stakeholders

Inundation depth at the grid level and the water level of the Mekong River at Can Tho are useful information and when considered together allow the various stakeholders to understand the interdependencies of interventions taken at the various spatial scales. Hence, this information will raise awareness amongst the various stakeholders to collaborate and to identify the challenges and opportunities of interventions taken at the delta level, city level, neighbourhood level and property level (Nguyen et al., 2019). The outcome will result in more inclusive and likely new perspectives on FRM in the area. These future flood risk information can be used by strategic planners to formulate delta policies and directives to guide strategic goal setting, future planning and implementation of effective flood mitigation measures in the Mekong Delta.

Similarly, users from the city administration can use future floodwater depth hazard information to ascertain the vulnerability of communities across socio-economic divides and threats to critical infrastructure. This future quantitative information can also be used in public outreach initiatives, such as "Swamped in Melbourne" (Bertram et al.,

2017), to create awareness among communities as informed citizenry would be willing to cooperate with other stakeholders toward reduction of flood risk.

Like in most other flooding contexts, floods in Can Tho affect the urban poor and economically weaker sections in the city (Chinh et al., 2016a; Chinh et al., 2016b). By overlaying demographical data on the computed flood damages, the impact and vulnerability of different social strata can be determined by the city planners. According to Chinh et al. (2016b), the average damage due to the year 2011 flood event at every household was about USD 333/-, whereas the average monthly income of a vulnerable household was USD 185/- . This means these households lose two months of income to floods and have to spend money on their economic recovery as well. *Inform* can be further updated by integrating grid-level demographic and socio-economic data to generate the community-specific economic impact and vulnerability maps without any additional inputs efforts from the user. Such information can help the city administrators to develop location-specific socio-economic interventions, such as providing grants, subsidies, or soft loans for the vulnerable low-income households to flood-proof their houses (Radhakrishnan et al., 2018a); and also to assess the ripple effect of interventions within the neighbourhood.

With *Inform* outputs, city administration officials in Can Tho would be able to combine flood hazard information with the socio-economic profile at those grids and ascertain the vulnerability at those grids (Garschagen, 2014). Vulnerability assessment can help the city administration draft targeted awareness campaigns and preparedness measures together with the local communities and disaster relief agencies such as National Red Cross societies. Upon availability of grid-level socio-economic data and infrastructure data, the *Inform* can be updated to generate acritical infrastructure index and vulnerability map without any additional inputs efforts from the user.

Assessing the effectiveness of FRM measures

Long term risk-informed perspective enables planners to explore and select adaptation pathways (sequences of measures in time) required to minimise flood risk for different scenarios. These measures encompass (amongst others) changing street profiles, land use planning and zoning based on recommendations such as EEA (2016).

The two interventions built into *Inform* represent the delta or basin scale action (i.e., reduction in river flow) and the local action at city scale or neighbourhood scale (i.e., arresting land subsidence).

5.4 Pilot testing and evaluation of *Inform*

Inform is aimed at supporting rapid flood risk assessment and facilitating a co-designing approach to explore, identify and select flood risk reduction measures with stakeholders' participation. In order to test and evaluate the tool, a multi-stakeholder co-design meeting was planned as part of the development process. However, due to travel restrictions associated with the Covid-19 pandemic, this co-design meeting in Can Tho could not be executed. Therefore, a pilot test of the tool was conducted with experts from universities and institutes in the Netherlands and Vietnam, and colleagues (PhD fellows at IHE Delft). These participants served as potential stakeholder to mimic the co-design work sessions and were asked to evaluate *Inform* based on the inputs and outputs generated within the framework of this study for Can Tho. The participants were asked to (i) evaluate the tool using seven criteria (the requirements see Section 5.2 and 5.3) and (ii) suggest improvements for each criterion?

Nine experts participated in the pilot testing, including six senior experts and three PhD researchers together covering a range of disciplines such as River Engineering and Water Governance, Hydrology and Water Resource Management, Hydraulic Engineering, Hydroinformatics, Numerical Ocean Modelling, etc.

Each expert received a user manual along with the tool and was requested to send their feedback in writing. The results are summarized in Table C.1 in Appendix C.

Based on the feedback, 4 out of 7 criteria received a positive evaluation from the participants. The remaining 3 criteria viz reliability of the tool, transcending coarser and finer resolution across spatial scales and, time scales could not be thoroughly assessed by the participants as these three parameters are at the backend of the tool - comprise the model and database that generates these results - is not evident to the users. As recommended by the expert participants the reliability of these criteria can be improved by adding additional information about the methodology, database and similar tools.

Based on the results of this pilot test, it can be cautiously concluded that *Inform* is a user-friendly interactive tool and easy to use even for non-water specialists. It satisfies general user requirements for an interactive tool that is suitable for co-designing work sessions. However, the latter needs to be further tested with a broader group of users with different expertise and different experience levels beyond the COVID-19 era in a multi-stakeholder face-to-face co-design setting.

Improvements suggested by the pilot users related to improved ease of use and scientific terminology, etc., will be implemented in the next version of the tool. In addition, the comment related to adding more reduction measures is discussed in detail in Section 5.5.

5.5 Discussion

5.5.1 Comparison with the recent study on the probabilistic flood risk calculation approach

The *Inform* approach aims at reducing the computational load during the process of flood risk calculation and adopts the same basic approach of the recently published FLORES (Berchum et al., 2020) for probabilistic flood risk calculations. However, the main point of departure in the *Inform* methodology is the strategic use of 1D and 2D hydraulic models. The simplification in *Inform* is achieved through the schematization of the 1D models at the delta level or basin level that can simulate the inputs – river water level based on upstream river flow and extreme sea level - for the computation of risk assessment in a matter of seconds. The pre-determined 2D models for fixed return periods serve as a base for interpolation of flood risk for the desired river water level. As the 1D model and coupled 1D-2D modelling results are used together in *Inform*, it is possible to assess the effectiveness of flood risk reduction strategies at a relatively fine resolution. The processes adopted by *Inform* are tailor-made to the context and requires careful schematization of the 1D model. Though it takes time to produce the simplified schematization, the comprehensiveness of computational results that can then be obtained from the 1D/2D flood model enables the detailed assessment of FRRS.

Compared to *Inform*, the FLORES approach which adopts simplified physical process descriptions, can be considered as a top-down methodology to rapidly screen a broad range of strategies and shortlist FRRS that can be further explored at finer resolution using the heavily context-based *Inform* to determine their technical feasibility. In effect, FLORES can be seen as a possible coarse screen for FRRS, and those strategies that pass through this sieve can be further fine sieved using *Inform* to assess risk reduction afforded by measures and provide inputs for cost benefit analyses.

5.5.2 Scope for improving *Inform* tool

Although *Inform* has been developed successfully with the must-have features of a co-design tool (e.g. inbuilt input library, flexible options, easy to use, quick results, user-friendly interface), there is still scope for improving *Inform* in the following aspects:

(1) Expanding the number of type of FRM interventions

The current version of *Inform* includes two types of interventions: intervention 1 - Reduction of flow due to upstream dams, intervention 2 – Land subsidence control in Can Tho city), which can be of use to all the stakeholders. *Inform* could be expanded to include local FRM interventions - either at city scale or/and locally (grid level) with soft structural measures, such as swales, wetlands, buffer areas; and hard structural measures, such as heightening dikes, constructing flood walls, pump station, relocation of critical infrastructure, changing street profiles and flood proofing of houses on the flood hazard, to enable the generation of flood depth and flood damage information. This will result in a powerful decision-making tool. The current version of *Inform* does not support such calculations as simulating these type of interventions (e.g. flood barriers, flood walls, etc.) is beyond the capacity of the 2D model. This will require that flood modellers and online tool developers jointly explore the possibility of incorporating a suite of FRM interventions such as Climateapp (Bosch Slabbers et al., 2014) in the flood risk tool. Combining past events and plausible future events can help in testing the effectiveness of FRM measures such as dikes, diversions, green infrastructure or nature-based solution, land use planning and zoning regulations.

Table 5.2 summarises possible local scale flood reduction measures that can potentially be applied in the model framework to increase the number of types of FRM interventions, in addition to the two interventions considered in the present version of *Inform*.

Table 5.2. Reduction measures at the local scale and how they can be implemented in the current model framework in future studies

Measure	Implementing in the model framework
1. Soft structural measures	
Swales, wetlands, buffer areas	Currently, there is no option for implementing these measures into the current model. However, it can be used in the same way as for the upstream dam construction by using reduction estimates in the flood inundation level when applying these solutions. Therefore, it is necessary to first investigate the effectiveness of these measures in reducing flood inundation levels at Ninh Kieu district.
2. Hard structural measures	
Heightening dikes	Adjusting the DEM data at the locations of dikes and then rerunning the 2D model to generate flood hazard and damage maps and calculating total damage corresponding to flood events.
Building reservoirs	Reservoirs can be implemented in the 1D/2D coupled model for the Ninh Kieu district as storages, and their capacity is described by a function or table of surface area versus height (Rossman, 2015).
Pumping stations	Pumps can be implemented in the 1D/2D coupled model for the Ninh Kieu district as the links to lift water to higher elevations. The relation between a pump's flow rate and

	conditions at its inlet and outlet nodes is described by a pump curve (Rossman, 2015).
New urban drainage system	A new drainage system can be added into the current 1D/2D coupled model with appropriate parameters (e.g. size and roughness coefficient of the pipes)

(2) Generating plausible future scenario particulars

Inform can be modified to generate plausible future scenarios. The output will then enable city planners and administrators to explore risk-informed FRM approaches based on a cost-benefit ratio assessment, instead of the standards-based approach, to decide on effective FRM interventions for the future. Also, total flood damages and grid-based flood damages can be calculated for different scenarios combinations of RCPs and MDPs. This will open up a wide range of analysis possibilities at the strategic, planning and project implementation level. A thorough exploration of flood damages and flood risk based on these scenarios and using this information to create engagement with stakeholders can lead to enhanced trust and awareness.

(3) Expanding objectives of flood risk assessment

Currently, *Inform* considers only fluvial flooding and does not include pluvial flooding. Water levels in Can Tho depend on the upstream river flow and downstream sea level. In which the upstream river flow is mainly attributed to the rainfall contribution in the upstream area. Therefore, modelling the delta wide rainfall contribution to the water level of Mekong may not be necessary. Besides, simulating the delta rainfall will lead to model expansion. As a result, simulation run-time will increase, and the model will no longer be sufficiently fast to support quantitative flood risk assessments. The inclusion of rainfall locally for Ninh Kieu district for simulating the flood depths and flood damages will improve the results of *Inform*. However, simulating the effect of local rainfall on inundation levels in the Ninh Kieu district is beyond the capability of the simplified 1D model, which is the main engine of *Inform*.

Additionally, there is evidence of floods affecting the health of people in Can Tho, which could also be explored from a public health point of view (Nguyen et al., 2017a).

Inform can be improved to assess the contribution of local rainfall on the flood hazard and risk in the urban centre of Can Tho, and the effect of flood hazards on public health and quantifying the associated health risk. This will result in a more robust and comprehensive tool to support flood risk assessment for the Ninh Kieu district in the future.

(4) Increasing the applicability and interactivity

Inform can be made compatible to be used through computer and smartphones alike, and it can serve as a good FRM outreach tool. *Inform* can become an integral part of flood risk awareness campaigns in Can Tho, which would lead to outcomes such as more informed autonomous adaptation approaches, align bottom-up and top-down approaches and reduce the build trust among the stakeholders.

There is conclusive evidence in cities of Can Tho, Da Nang and Quy Nhon in Vietnam that people resort to coping measures to avoid damages during seasonal flooding events; their intention to implement adaptation measures to minimise or eliminate long term flood risk is moderate; and, there is a willingness to learn and adapt to long term climate impacts (Ngo et al., 2020a). Tools such as *Inform* can enable such learning and promote the implementation of adaptation measures.

5.6 Conclusions

Inform, a web-based hydraulic tool was developed based on a simplified 1D model for the entire Mekong Delta, flood hazard and damage maps, and estimated flood damages for Ninh Kieu district to support FRM in Ninh Kieu district in the future. Seven criteria extracted from a literature review were used for reference as guidance during the development of *Inform*. A first pilot test of the tool revealed that users perceived *Inform* to contain features required in a co-designing tool (e.g. inbuilt input library, flexible options, easy to use, providing quick results, user-friendly interface), indicating that *Inform* is a promising interactive tool that could support probabilistic flood risk assessment and facilitate a co-designing approach for risk reduction measures with the participation of multi-stakeholders.

Inform has been shown to have the potential to help stakeholders to better understand and reflect on the flood risk they are exposed to. Risk perceptions allow the prediction of individual and community responses, enable policymakers to develop effective strategies, implement risk management interventions that are in line with public expectations, and encourage dialogue and collective learning among stakeholders.

Inform may also be useful in agenda-setting, formulation of plans and implementation of urban FRM objectives in the Mekong Delta. The instant flood risk modelling tool – *Inform* – presented here can support the identification and selection of interventions that leads to sustainable infrastructure delivery. Further pilot testing is recommended with a wider group of stakeholders *on the ground* to substantiate the abovementioned preliminary conclusions.

Chapter 6

General Conclusions

6.1 Introduction

The preceding chapters described probabilistic flood forecasting and risk assessment (flood hazard and risk) for the urban centre of a coastal city - Can Tho city, Mekong Delta, Vietnam for present-day and future, and also presented a co-designing interactive tool aimed at supporting the rapid assessment of flood risk to inform the determination of economically and socially acceptable flood risk reduction measures. This chapter presents the general conclusions of the study. Section 6.3 outlines the limitations associated with this study and provides suggestions for future research.

6.2 General conclusions

The successful application of the 1D SWMM model to the Mekong Delta, forced with both upstream riverflow and downstream extreme sea levels (tide + storm surge) in this study demonstrated that it is possible to simulate river water levels, with an acceptable level of the accuracy, at a location of interest in a complex, deltaic river system, such as the lower Mekong with a relatively simple (simplified from thousands to several tens of cross-sections) and fast (reduction of simulation time from 1.5 h to around one minute, for a 1 year simulation) 1D hydraulic model. With these features, such a simplified model is a feasible solution for probabilistic flood risk assessment and for stakeholder-based co-design applications, which require running thousands of model simulations for a given scenario. Additionally, the application of the SWMM model, a model originally developed to simulate drainage/sewerage systems for large river simulations is a unique feature of this study. As SWMM is an open-source model and simple to use, it is ideal option for multi-stakeholder co-designing efforts.

Combining the above mentioned simplified 1D hydrodynamic model for the entire Mekong Delta with a detailed 1D/2D coupled model, an efficient (i.e. fast, accurate and low computational cost) approach to quantitatively assess flood risk was established for Ninh Kieu district, the urban centre of Can Tho city. Key features of the modelling approach include (a) Model reduction and compound forcing - a substantially simplified 1D model for the entire Mekong Delta (area of 40,577 km^2) which can simulate one year of river water levels with concurrent upstream discharge and downstream extreme sea levels (with an hourly time step) in under 60 seconds (Chapter 2), (b) strategic use of river water level estimates to drive a detailed 1D/2D hydrodynamic flood inundation model (1D/2D model) that is focused on the area of interest (the urban centre of Can Tho city), (c) reduction of the number of 1D/2D coupled model runs required by performing flood frequency analysis based on flood hydrograph patterns, (d) derivation of probabilistic flood hazard maps for the present and in 2050 under RCP 4.5 and 8.5, with and without land subsidence; and (e) quantification of flood risk for the present and in 2050 under RCP 4.5 and 8.5, with and without land subsidence.

Probabilistic flood hazard maps for Ninh Kieu district, the urban centre of Can Tho city, indicated that even under the current situation, more than 12 % of the study area would be inundated by the 100 yr return period of water level. With climate change, but without land subsidence, the 100 yr return period flood extent is projected to more than double by 2050, with not much of difference between the two climate scenarios considered (RCP 4.5 and RCP 8.5). However, if the present rate of land subsidence will continue in the future the 0.5 yr and 100 yr return period flood extents are projected to increase by around 15-fold and 8-fold by 2050 under RCP 4.5 and RCP 8.5 respectively (relative to the present-day flood extent that would result from the same return period water level). The projected 15-fold increase in flood extent projected by 2050 for the twice per year (0.5 yr return period) flood event in particular implies that the "do nothing" management approach is not a feasible option for Can Tho.

Flood risk assessment results indicated that the current flood risk of the Ninh Kieu district, expressed as EAD, is 5.3 million USD/year. The EAD in 2050 under RCP 4.5 and RCP 8.5 will increase by 1.7 times and 1.8 times compared to present-day. Land subsidence has a more significant effect on the EAD than the impact of climate change, resulting in increases (relative to present-day) of approximately 20-fold and 21-fold by 2050 under RCP 4.5 and RCP 8.5.

The instant flood risk modelling tool – *Inform*, was developed based on the simplified 1D model for the entire Mekong Delta, flood hazard and damage maps, and estimated flood damage. In addition, seven requirements extracted from a literature review were used for reference as guidance during the development of the tool. Pilot testing with experts indicated that *Inform* contains the must-have features of a co-design tool (e.g. inbuilt input library, flexible options, easy to use, quick results, user-friendly interface), providing confidence in its potential as an interactive tool that can provide rapid flood risk assessments with quantitative information (e.g. flood levels, flood hazard and damage maps, estimated damages) required for co-designing efforts aimed at flood risk reduction.

Inform can help stakeholders to have a better understanding of the flood risk they are exposed to and encourage them to reflect on possible consequences. Risk perceptions allow to predict individual and community responses, enable policymakers to develop effective strategies, implement risk management interventions that are in line with public expectations, and encourage dialogue and collective learning among stakeholders. The instant flood risk modelling tool – *Inform* – presented here can be seen as a tool that has the potential to support successful implementation of interventions that leads to sustainable infrastructure delivery.

The flood modelling system approach presented in this study for probabilistic fluvial flood forecasting and risk assessment for Can Tho city, Mekong Delta, Vietnam has proven itself to be a rapid approach with low computational cost. Apart from the input data, as the approach does not contain any site specificities, it can be replicated in other areas in the Mekong Delta, and even in other parts of the world for flood risk assessment. This approach would especially be useful in the developing countries facing limitations in the availability of good quality hydrological and economic data and in computational resources.

6.3 Limitations of the study and possible future research initiatives

In addition to the usefulness of this approach for probabilistic fluvial flood forecasting and risk assessment was mentioned above, the simplified approach adopted here contains some limitations. This section identifies the limitations that are directly related to the modelling approach, and provides future research directions that could help to overcome or circumvent them.

The use of the simplified 1D model is appropriate for the purpose of this study. It provides rapid and accurate estimates of water level at a location of interest (here, Can Tho city). However, the accuracy of the predicted water levels for other locations along the river (Chau Doc and Tan Chau) is not optimal. This is related to the level of detail in the model, which was reduced significantly from thousands to several tens of cross-sections. With such a limited number of cross-sections, it is impossible to accurately represent the river cross-section at all locations, which is one of the main factors that govern the computed water level. The accuracy of the simplified model results at locations along the river can be improved if the level of detail of the model is increased. However, this will lead to an increase the simulation run-time, and as a result, may compromise the main purpose of this study. Future research in this area could focus on providing more accurate predictions at multiple locations by increasing the level of detail only around the locations of interest while still ensuring sufficiently low computational cost. If this is achieved, it will facilitate large scale (e.g. catchment scale) flood risk assessment with low computational costs.

The projected changes in sea level (tide + surge) at the Mekong River mouths in the future did not consider potential climate change driven variations in storm surge. Taking into account future changes in storm surges may lead to an increase in predicted river water levels at Can Tho, and further an increase in the future flood hazard and risk, which will improve flood risk assessment for the study area.

The use of stage-damage functions that were developed for Ho Chi Minh City to calculate the flood damage and risk for Can Tho is another limitation of this study, even though there are many similarities between the two cities as mentioned in Chapter 4. Stage-damage functions were developed by Lasage et al. (2014) for different land use categories in District 4 of Ho Chi Minh City through household surveys, taking into account typologies of assets such as buildings, furniture and roads. In comparison, the use of satellite imagery (in Google Maps) to determine the typology distribution of assets within the land-use classes in Ninh Kieu district is relatively rudimentary in this study. The use of stage-damage functions for Ho Chi Minh City together with the less accurate land-use information in Ninh Kieu district is likely to lead to less accurate computed flood damage and EAD values in Ninh Kieu district in Chapter 4. However, as mentioned in Chapter 4, since stage-damage functions for different typologies of assets are currently not available for Ninh Kieu, this is the best that can currently be done in terms estimating flood damages and EADs in Can Tho. Future research in this area could benefit from developing stage-damage functions for the respective typologies of assets specifically in Ninh Kieu district through household surveys.

The gaming tool *Inform,* which was aimed at supporting a co-designing approach for risk reduction measures in the urban centre of Can Tho city (Ninh Kieu district) among stakeholders, has not yet been used on-site (Ninh Kieu district) as planned, due to travel restrictions associated with the Covid-19 pandemic. In addition, the current reduction measures used in *Inform* are limited and the outputs have not been yet used to perform cost/benefit analyses. Expansion of the array of possible reduction measures (e.g. heightening dikes, constructing pumping stations, reservoirs, and new urban drainage systems) that can be implemented in *Inform* and adding a feature to calculate cost/benefit for each measure in further developments of the tool would improve its capabilities as a decision-making tool. Moreover, assessment of the accuracy and features of the tool based on the experience and input of stakeholders (experts, technicians, and local inhabitants) would be invaluable to fine-tune and further develop *Inform.*

Flood risk assessment for Ninh Kieu district, taking into account the effects of climate change and land subsidence was not assessed beyond 2050 in this study. This is due to a lack of robust regional scale climate change forcing (e.g. river flow, sea-level rise and storm surge) projections for the study area beyond 2050. When such robust projections are available for the Mekong Delta, one useful future research direction would be to extend this analysis till the end of the 21st century, which would assist in long term planning of adaptation strategies.

The contribution of local rainfall on the flood hazard and risk in the urban centre of Can Tho city (Ninh Kieu district) was not investigated in this study. Inclusion of this effect may influence the flood extent and inundation depth at Ninh Kieu district (Apel et al., 2016). Therefore, a future research direction would be to assess flood hazard and risk assessment for Ninh Kieu district due to the combination of fluvial and pluvial floods, taking into account the effect of climate change on local rainfall, river flow and sea level, as well as land subsidence. This will provide better estimates of the potential flood hazards and risks in Ninh Kieu district in the future.

There is evidence of floods affecting the health of people in Can Tho, which could also be explored from a public health point of view (Nguyen et al., 2017a). This is because the floodwater quality at many locations in Can Tho is significantly affected by contamination with pollutants from sewage (Nguyen et al., 2017b). Future research in this area could benefit from assessing the effect of flood hazards on public health and quantifying the associated health risk.

'Living with floods' is an emerging concept in many parts of the world, especially in the Mekong Delta. At the same time, the effects of climate change and land subsidence will increase the flood hazard and exacerbate the flood damages in the study area as shown in this study. This in turn may lead to changes in the perception and appraisal of inhabitants of the concept 'Living with floods' in the future. Yet another direction of future research is to combine the results of this study with agent-based models to capture the individuals' behaviour to analyse the change in flood risk perceptions related to the effects of climate change and land subsidence.

Appendix

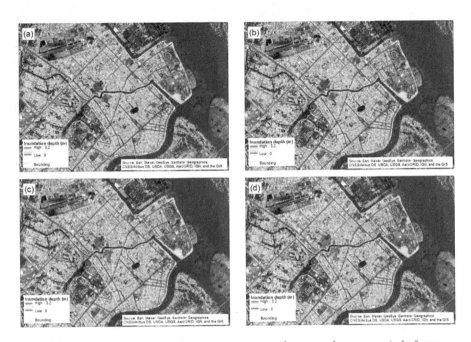

Figure A.1. Flood hazard maps for the present corresponding to each return period of water level, (a) 1 yr return period, (b) 2 yr return period, (c) 10 yr return period, (d) 20 yr return period

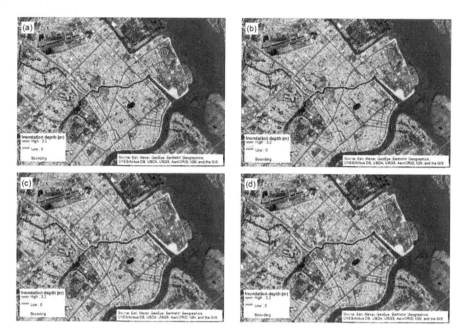

Figure A.2. Flood hazard maps for 2050 under RCP 4.5 (model scenario #2) corresponding to each return period of water level, (a) 1 yr return period, (b) 2 yr return period, (c) 10 yr return period, (d) 20 yr return period

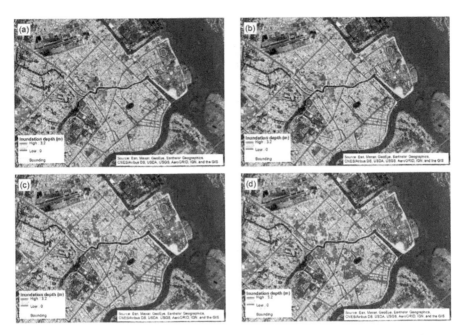

Figure A.3. Flood hazard maps for 2050 under RCP 8.5 (model scenario #4) corresponding to each return period of water level, (a) 1 yr return period, (b) 2 yr return period, (c) 10 yr return period, (d) 20 yr return period

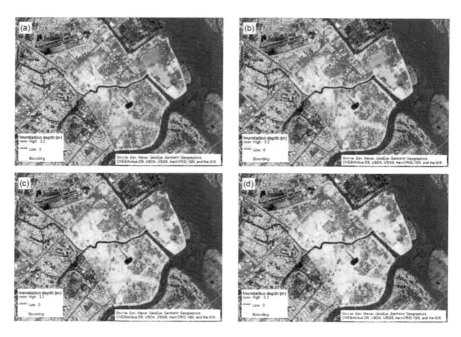

Figure A.4. Flood hazard maps for 2050 under RCP 4.5 (model scenario #1) corresponding to each return period of water level, (a) 1 yr return period, (b) 2 yr return period, (c) 10 yr return period, (d) 20 yr return period

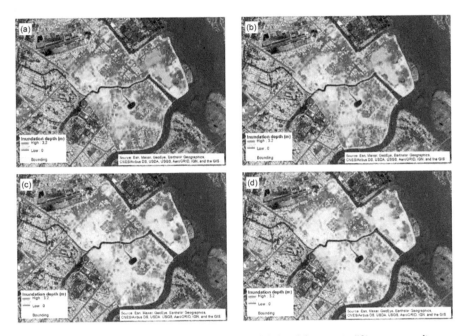

Figure A.5. Flood hazard maps for 2050 under RCP 8.5 (model scenario #3) corresponding to each return period of water level, (a) 1 yr return period, (b) 2 yr return period, (c) 10 yr return period, (d) 20 yr return period

119

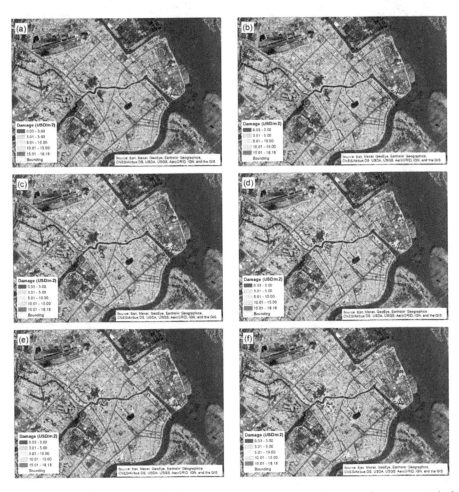

Figure B. 1. Flood damage maps for baseline situations corresponding to each return period of water level at Can Tho, (a) 1 yr return period, (b) 2 yr return period, (c) 5 yr return period, (d) 10 yr return period, (e) 20 yr return period, (f) 50 yr return period

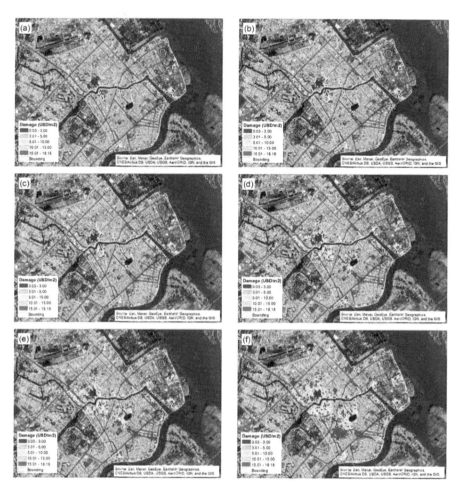

Figure B.2. Flood damage maps for 2050 (scenario #1) corresponding to each return period of water level at Can Tho, (a) 1 yr return period, (b) 2 yr return period, (c) 5 yr return period, (d) 10 yr return period, (e) 20 yr return period, (f) 50 yr return period

122

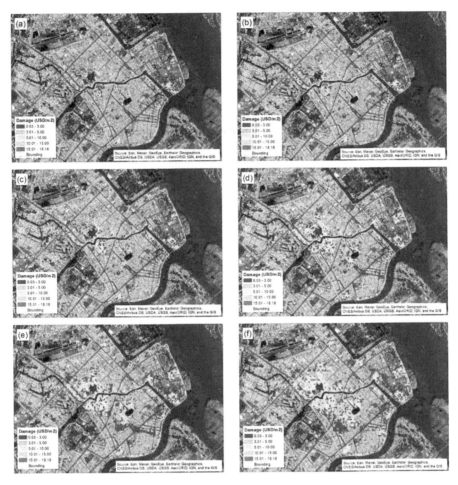

Figure B.3. Flood damage maps for 2050 (scenario #3) corresponding to each return period of water level at Can Tho, (a) 1 yr return period, (b) 2 yr return period, (c) 5 yr return period, (d) 10 yr return period, (e) 20 yr return period, (f) 50 yr return period

123

Table B.1. Total damage and average annual damage corresponding to each return period for the present and for 2050 (under RCP 4.5 and RCP 8.5) without land subsidence, and the EAD for each scenario

Return period T (year)	Pi	ΔPi	Present		2050 (RCP4.5)		2050 (RCP4.5)	
			Total damage (Mil. USD)	Average annual damage (Mil. USD)	Total damage (Mil. USD)	Average annual damage (Mil. USD)	Total damage (Mil. USD)	Average annual damage (Mil. USD)
0.5	2.000		2.38		3.49		3.64	
		1.000		2.46		3.91		3.76
1	1.000		2.53		4.02		4.18	
		0.500		1.30		2.21		2.20
2	0.500		2.67		4.66		4.77	
		0.300		0.86		1.59		1.64
5	0.200		3.03		5.93		6.17	
		0.100		0.32		0.64		0.67
10	0.100		3.31		6.89		7.30	
		0.050		0.18		0.37		0.41
20	0.050		3.69		8.06		9.24	
		0.030		0.12		0.28		0.33
50	0.020		4.52		10.82		12.67	
		0.010		0.05		0.12		0.14
100	0.010		5.30		12.97		15.39	
Expected Annual Damage (EAD)				**5.27**		**8.93**		**9.34**

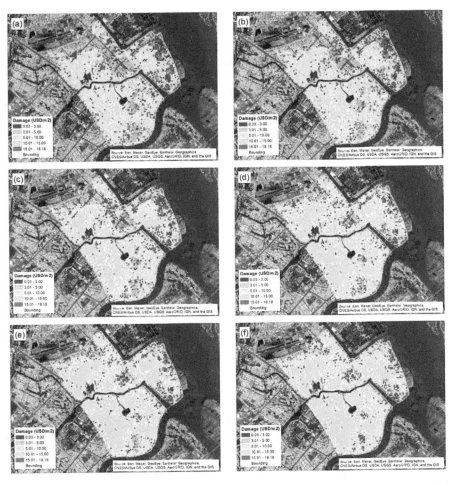

Figure B.4. Flood damage maps for 2050 (scenario #1) corresponding to each return period of water level at Can Tho, (a) 1 yr return period, (b) 2 yr return period, (c) 5 yr return period, (d) 10 yr return period, (e) 20 yr return period, (f) 50 yr return period

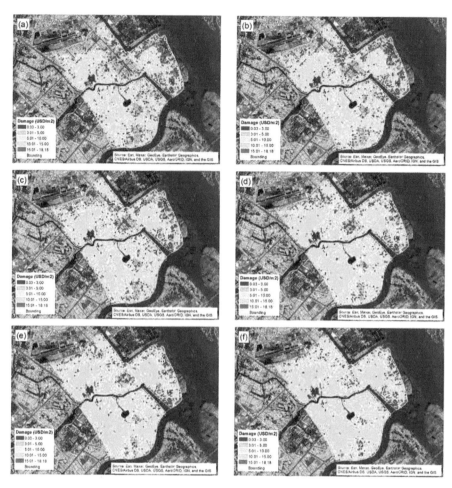

Figure B.5. Flood damage maps for 2050 (scenario #3) corresponding to each return period of water level at Can Tho, (a) 1 yr return period, (b) 2 yr return period, (c) 5 yr return period, (d) 10 yr return period, (e) 20 yr return period, (f) 50 yr return period

Table B.2. Total damage and annual damage corresponding to each return period for 2050 (under RCP 4.5 and RCP 8.5) with land subsidence, and expected annual damage for each scenario

Return period T (year)	Pi	ΔPi	2050 (RCP4.5+ Land subsidence)		2050 (RCP8.5+ Land subsidence)	
			Total damage (Mil. USD)	Average annual damage (Mil. USD)	Total damage (Mil. USD)	Average annual damage (Mil. USD)
0.5	2.00		46.35		47.94	
		1.00		48.74		50.39
1	1.00		51.13		52.84	
		0.50		26.44		27.56
2	0.50		54.61		57.39	
		0.30		17.17		17.98
5	0.20		59.83		62.50	
		0.10		6.13		6.39
10	0.10		62.75		65.21	
		0.05		3.21		3.35
20	0.05		65.49		68.65	
		0.03		2.03		2.11
50	0.02		69.71		72.31	
		0.01		0.71		0.73
100	0.01		71.99		74.18	
Expected Annual Damage (EAD)			**104.41**		**108.51**	

Appendix C

Table C.1. Summary of participants' feedback and suggestions for improving the Inform tool

No	Criteria	Participants' feedback	Suggestions to improve the tool
1	Ensuring reliability of tool outputs	- 7 participants said that it is difficult to evaluate the tool outputs' reliability due to lacking information related to the input data, calibration and validation of the tool. - 2 participants said that the tool's outputs are coherent with the selected input data.	- Adding a link/info box with information for users (e.g. the origin of the input parameters (discharge, sea level), the tool calibration and validation). - Adding a comparison feature allows the user to see the output of two consecutive selections at once instead of resetting the system. - Increasing the size and resolution of maps.
2	Ease of use and avoiding input overkill	- 7 participants said that the tool is easy to use with the friendly-user interface. - 1 participant said that it is moderate. - 1 participant did not give a specific evaluation.	- Collecting more recent data (discharge, sea level). - Adding units of flow, sea level in the respective graphs. - Adjusting the tool to ensure that when selecting an input parameter (e.g. discharge) for a specific year the other one (e.g. sea level) should change automatically for the same year. - Creating three or four pre-programmed scenarios. - Creating an additional Vietnamese version of the tool
3	Time taken to generate tool outputs	All participants said that the tool is rapid in generating the outputs.	
4	Transcending coarser and finer resolutions across spatial scales	- 3 participants said that the spatial scales is fine. - 2 participants said that it was a lack of information for evaluation. - 4 participant did not give a specific evaluation.	- Adding needed information to the user about this criterion, including the meaning and purpose of this criterion.

5	Transcending coarser and finer resolutions across temporal scales	- 2 participants said that transcending resolution across temporal scales is fine. - 1 participants said that it was a lack of information for evaluation. - 6 participant did not give a specific evaluation.	- Adding needed information to the user about this criterion, including the meaning and purpose of this criterion. - Analyzing all flood events from 2000 to 2009 in terms of hazard to select some typical flood events to include in the tool instead of using all flood events in the database and also, providing descriptions of these typical flood events.
6	Interpretation and relevance of tool outputs across a wide spectrum of stakeholders	- 7 participants said that this tool provides useful outputs for stakeholders. - 2 participant did not evaluate this criterion	- Adjusting scientific terminology used as below: + Upstream boundary condition should be replaced by "river discharge". + Sea level distribution should be replaced by "Sea level". + Phase change should be replaced by "Initial time".
7	Assessing the effectiveness of FRM measures	- 7 participants said that the effectiveness of FRM measures is good. - 2 participants did not evaluate this criterion.	- "Interventions" should be a separate section and it measures should be visible instead of hidden. - Adding more reduction measures into Interventions section (e.g. flood walls, dykes, reservoirs, water tanks, pump stations, etc.).

References

Adaptation Sub-Committee.: Climate Change—is the UK preparing for flooding and water scarcity? Committee on Climate Change, London, 2012.

Adeogun, A. G., Daramola, M. O., and Pathirana, A.: Coupled 1D-2D hydrodynamic inundation model for sewer overflow: Influence of modeling parameters, Water Sci., 29(2), 146–155, doi:10.1016/j.wsj.2015.12.001, 2015.

Adnan, M.S.G., Abdullah, A.Y.M., and Dewan, A.; Hall, J.W.: The effects of changing land use and flood hazard on poverty in coastal Bangladesh, Land use policy., 99, 104868, doi:10.1016/j.landusepol.2020.104868, 2020.

Aerts, J. C. J. H., Lin, N., Botzen, W., Emanuel, K., and de Moel, H.: Low-Probability Flood Risk Modeling for New York City, Risk Anal., 33(5), 772–788, doi:10.1111/risa.12008, 2013.

Akpo, E., Crane, T.A., Vissoh, P.V., and Tossou, R.C.: Co-production of knowledge in multi-stakeholder processes: Analyzing joint experimentation as social learning. J. Agric. Educ. Ext, 21, 369–388, doi:10.1080/1389224X.2014.939201, 2015.

Alfieri, L., Burek, P., Dutra, E., Krzeminski, B., Muraro, D., Thielen, J., and Pappenberger, F.: GloFAS – global ensemble streamflow forecasting and flood early warning, Hydrol. Earth Syst. Sci., 17, 1161–1175, doi:10.5194/hess-17-1161-2013, 2013.

Alfieri, L., Burek, P., Feyen, L., and Forzieri, G.: Global warming increases the frequency of river floods in Europe, Hydrol. Earth Syst. Sci., 19, 2247–2260, https://doi.org/10.5194/hess-19-2247-2015, 2015.

Alfieri, L., Bisselink, B., Dottori, F., Naumann, G., Wyser, K., Feyen, L., and Roo, A. De.: Future Global projections of river flood risk in a warmer world, Earth's Future, doi:10.1002/2016EF000485, 2017.

Apel, H., Aronica, G.T., Kreibich, H and Thieken, A.H.: Flood risk analyses–how detailed do we need to be?, Nat Hazards 49, 79–98, https://doi.org/10.1007/s11069-008-9277-8, 2009.

Apel, H., Trepat, O. M., Hung, N. N., Chinh, D. T., Merz, B. and Dung, N. V.: Combined fluvial and pluvial urban flood hazard analysis: concept development and application to Can Tho city, Mekong Delta, Vietnam, Nat. Hazards Earth Syst. Sci., 941–961, doi:10.5194/nhess-16-941-2016, 2016.

Arkema, K. K., Guannel, G., Verutes, G., Wood, S. A., Guerry, A., Ruckelshaus, M., Kareiva, P., Lacayo, M. and Silver, J. M.: Coastal habitats shield people and property from sea-level rise and storms, Nature Climate Change, 3(10), 913–918, doi:10.1038/nclimate1944, 2013.

Arnell, N. W., and Gosling, S. N.: The impacts of climate change on river flood risk at the global scale, Clim. Change., 134, 387–401, doi:10.1007/s10584-014-1084-5, 2016.

Aubert, A. H., Bauer, R., and Lienert, J.: A review of water-related serious games to specify use in environmental Multi-Criteria Decision Analysis. Environ Modell Softw., 105, 64-78, doi:https://doi.org/10.1016/j.envsoft.2018.03.023, 2018.

Balica, S.F., Wright, N.G., and van der Meulen, F.: A flood vulnerability index for coastal cities and its use in assessing climate change impacts, Nat Hazards., 64, 73–105, doi:10.1007/s11069-012-0234-1, 2012.

Balica, S.F., Dinh, Q., Popescu, I., Vo, T.Q., and Pham, D.Q.: Flood impact in the Mekong Delta, Vietnam, J. Maps., 10, 257–268, doi:10.1080/17445647.2013.859636, 2014.

Bates, P.D., and De Roo, A.P.J.: A simple raster-based model for flood inundation simulation, J. Hydrol., 236(1-2), 54-77, 2000.

Beall, A., and Zeoli, L.: Participatory modeling of endangered wildlife systems: Simulating the sage-grouse and land use in Central Washington, Ecol Econ., 68(1–2), 24–33, doi:10.1016/j.ecolecon.2008.08.019, 2008.

Berends, K. D., Warmink, J. J. and Hulscher, S. J. M. H.: Efficient uncertainty quantification for impact analysis of human interventions in rivers, Environmental Modelling and Software, 107(September), 50–58, doi:10.1016/j.envsoft.2018.05.021, 2018.

Berends, K., Straatsma, M., Warmink, J. and Hulscher, S.: Uncertainty quantification of flood mitigation predictions and implications for decision making, Uncertainty quantification of flood mitigation predictions and implications for decision making, 1–25, doi:10.5194/nhess-2018-325, 2018.

Bertram, N., Murphy, C., Pasman, R., Rogers, B., Gunn, A., Urich, C., Arnbjerg-Nielsen, K., Lowe, R., Radhakrishnan, M., and Gersonius, B.: Swamped, Swamped - The Gallery, Victoria, Australia, 22 Feb 2017 to 22 Mar 2017, 2017.

Bezak, N., Brilly, M., and Šraj, M.: Comparison between the peaks-over-threshold method and the annual maximum method for flood frequency analysis, Hydrol. Sci. J., 59(5), 959–977, doi:10.1080/02626667.2013.831174, 2014.

Birkmann, J., Garschagen, M., Kraas, F. and Quang, N.: Adaptive urban governance: New challenges for the second generation of urban adaptation strategies to climate change, Sustainability Science, 5(2), 185–206, doi:10.1007/s11625-010-0111-3, 2010.

Bomers, A., Schielen, R. M. J. and Hulscher, S. J. M. H.: Application of a lower-fidelity surrogate hydraulic model for historic flood reconstruction, Environmental Modelling and Software, 117(January 2018), 223–236, doi:10.1016/j.envsoft.2019.03.019, 2019a.

Bomers, A., van der Meulen, B., Schielen, R. M. J. and Hulscher, S. J. M. H.: Historic Flood Reconstruction With the Use of an Artificial Neural Network, Water Resources Research, 55(11), 9673–9688, doi:10.1029/2019WR025656, 2019b.

Bomers, A., Schielen, R. M. J. and Hulscher, S. J. M. H.: Decreasing uncertainty in flood frequency analyses by including historic flood events in an efficient bootstrap approach, Natural Hazards and Earth System Sciences, 19(8), 1895–1908, doi:10.5194/nhess-19-1895-2019, 2019c.

Bosch Slabbers, Deltares, Sweco, Witteveen-Bos, & KNMI. (2014). Climate adaptation app: https://www.climateapp.nl/, last access: 6 August 2020.

Brown, C., Alexander, P., Holzhauer, S., Rounsevell, M.D.A.: Behavioral models of climate change adaptation and mitigation in land-based sectors, Wiley Interdiscip Rev Clim Chang., 8(2):1757–7799, 2017.

Brügger, A., Morton, T. A., and Dessai, S.: "Proximising" climate change reconsidered: A construal level theory perspective, J. Environ. Psychol., 46, 125-142, doi:10.1016/j.jenvp.2016.04.004, 2016.

Bureau Reclamation: https://www.usbr.gov/ssle/damsafety/risk/methodology.html, last access: 11 May, 2020.

Can Tho City People's Committee.: Can Tho City Climate Change Resilience Plan, Can Tho City People's Committee, Can Tho, Vietnam, 30 pp., 2010.

Cardoso, M. A., Almeida, M. C., Brito, R. S., Gomes, J. L., Beceiro, P., and Oliveira, A.: 1D/2D stormwater modelling to support urban flood risk management in estuarine areas: Hazard assessment in the Dafundo case study, J. Flood Risk Manag., 1–15, doi:10.1111/jfr3.12663, 2020.

Chadwick, R., Ackerley, D., Ogura, T. and Dommenget, D.: Separating the Influences of Land Warming, the Direct CO_2 Effect, the Plant Physiological Effect, and SST Warming on Regional Precipitation Changes, Journal of Geophysical Research: Atmospheres, 124(2), 624–640, doi:10.1029/2018JD029423, 2019.

CCCO and ISET.: Peri-Urban Development Planning and Flooding Problems: Story of New Urban Areas in Can Tho City, Viet Nam, ISET, Hanoi, Vietnam, 2015.

Chen, R., Zhang, Y., Xu, D., Liu, M.: Climate Change and Coastal Megacities: Disaster Risk Assessment and Responses in Shanghai City, in: Climate Change, Extreme Events and Disaster Risk Reduction, Sustainable Development Goals Series, edited by: Mal, S., Singh, R., Huggel, C., Springer, Cham., doi:10.1007/978-3-319-56469-2_14, 2018.

Chinh, D., Gain, A., Dung, N., Haase, D., and Kreibich, H.: Multi-Variate Analyses of Flood Loss in Can Tho City, Mekong Delta, Water., 8(1), 6, doi:10.3390/w8010006, 2016a.

Chinh, D. T., Dung, N. V., Kreibich, H., and Bubeck, P.: The 2011 flood event in the Mekong Delta: preparedness, response, damage and recovery of private households and small businesses, Disasters., 753-78, doi:10.1111/disa.12171, 2016b.

Chinh, D. T., Dung, N. V., Gain, A. K., and Kreibich, H.: Flood Loss Models and Risk Analysis for Private, Water., 9, 313, doi:10.3390/w9050313, 2017.

Custer, R.: Hierarchical Modelling of Flood Risk for Engineering Decision Analysis, PhD thesis, Technical University of Denmark, Denmark, 212 pp, 2015.

Dasallas, L., Kim, Y., and An, H.: Case study of HEC-RAS 1D-2D coupling simulation: 2002 Baeksan flood event in Korea, Water., 11(10), 1–14, doi:10.3390/w11102048, 2019.

De Moel, H., Bouwer, L. M. and Aerts, J. C. J. H.: Uncertainty and sensitivity of flood risk calculations for a dike ring in the south of the Netherlands, Science of the Total Environment, 473–474, 224–234, doi:10.1016/j.scitotenv.2013.12.015, 2014.

De Moel, H., Jongman, B., Kreibich, H. and Merz, B.: Flood risk assessments at different spatial scales, Mitig Adapt Strateg Glob Change., 20, 865–890, doi:10.1007/s11027-015-9654-z, 2015.

Deltares.: Aqueduct global flood analyzer, https://www.deltares.nl/en/software/aqueduct-global-flood-analyzer/, last access: 25 September 2020.

den Haan, R. J., van der Voort, M. C., Baart, F., Berends, K. D., van den Berg, M. C., Straatsma, M. W., Geenen, A. J. P. and Hulscher, S. J. M. H.: The Virtual River Game: Gaming using models to collaboratively explore river management complexity, Environmental Modelling and Software, 134(August), 104855, doi:10.1016/j.envsoft.2020.104855, 2020

Dinh, Q., Balica, S., Popescu, I., and Jonoski, A.: Climate change impact on flood hazard, vulnerability and risk of the Long Xuyen Quadrangle in the Mekong Delta Climate change impact on flood hazard, vulnerability and risk of the Long Xuyen, Int. J. River Basin Manag., 10, 103–120, doi:10.1080/15715124.2012.663383, 2012.

Domeneghetti, A., Castellarin, A., and Brath, A.: Assessing ratingcurve uncertainty and its effects on hydraulic model calibration, Hydrol. Earth Syst. Sci., 16, 1191–1202, doi:10.5194/hess-16-1191-2012, 2012.

Dung, N. V., Merz, B., Bárdossy, A., Thang, T. D. and Apel, H.: Multi-objective automatic calibration of hydrodynamic models utilizing inundation maps and gauge data, Hydrology and Earth System Sciences, 15(4), 1339–1354, doi:10.5194/hess-15-1339-2011, 2011.

Dung, N. V., Merz, B., Bárdossy, A. and Apel, H.: Handling uncertainty in bivariate quantile estimation - An application to flood hazard analysis in the Mekong Delta, Journal of Hydrology, 527, 704–717, doi:10.1016/j.jhydrol.2015.05.033, 2015.

Dupuits, E. J. C., Diermanse, F. L. M. and Kok, M.: Economically optimal safety targets for interdependent flood defences in a graph-based approach with an efficient evaluation of expected annual damage estimates, Natural Hazards and Earth System Sciences, 17(11), 1893–1906, doi:10.5194/nhess-17-1893-2017, 2017.

DWF.: Survey on Perception of risk in Can Tho City, Lauzerte, France, 182 pp, 2011.

Eastham, J., Mpelasoka, F., Ticehurst, C., Dyce, P., Ali, R. and Kirby, M.: Mekong River Basin Water Resources Assessment: Impacts of Climate Change. In CSIRO Water for a Healthy Country National Research Flagship Report; CSIRO: Canberra, Australia, 153 pp, 2008.

Edelenbos, J., Van Buuren, A., Roth, D., and Winnubst, M.: Stakeholder initiatives in flood risk management: exploring the role and impact of bottom-up initiatives in three 'Room for the River' projects in the Netherlands, J. Environ. Plan. Manag., 60(1), 47–66, doi:10.1080/09640568.2016.1140025, 2017.

EEA.: Urban adaptation to climate change in Europe: Transforming Cities in a changing climate, Denmark, 135 pp, 2016.

Erban, L. E., Gorelick, S. M., and Zebker, H. A.: Groundwater extraction, land subsidence, and sea-level rise in the Mekong Delta, Vietnam, Environ. Res. Lett., 9, doi:10.1088/1748-9326/9/8/084010, 2014.

Erkens, G., Bucx, T., Dam, R., De Lange, G., and Lambert, J.: Sinking coastal cities, Proceedings of the International Association of Hydrological Sciences, 372(November), 189–198, doi:10.5194/piahs-372-189-2015, 2015.

Evans, E., Ashley, R., Hall, J., Penning-Rowsell, E., Saul, A., Sayers P., Thorne, C., Watkinson, A.: Foresight. Future Flooding. Scientific Summary: Volumes I and II. Office of Science and Technology, London, 2004.

Fan, Y., Ao, T., Yu, H., Huang, G., and Li, X.: A coupled 1D-2D hydrodynamic model for urban flood inundation, Adv. Meteorol., doi:10.1155/2017/2819308, 2017.

Faulkner, H., Parker, D., Green, C. and Beven, K.: Developing a translational discourse to communicate uncertainty in flood risk between science and the practitioner, Ambio, 36(8), 692–703, doi:10.1579/0044-7447(2007)36[692:DATDTC]2.0.CO;2, 2007.

FLOODsite.: Flood risk assessment and flood risk management. An introduction and guidance based on experiences and findings of FLOODsite. FLOODsite Deltares, Delft Hydraul. Delft, Netherlands, 143 pp, doi:9789081406710, 2009.

Forzieri, G., Cescatti, A., Batista, F., and Feyen, L.: Increasing risk over time of weather-related hazards to the European population: a data-driven prognostic study, Lancet Planet Health., 1(5), e200–e208, doi:10.1016/S2542-5196(17)30082-7, 2017.

Foudi, S., Osés-eraso, N., and Tamayo, I.: Integrated spatial flood risk assessment : The case of Zaragoza Land Use Policy Integrated spatial flood risk assessment, 42, 278–292, doi:10.1016/j.landusepol.2014.08.002, 2015.

Garschagen, M.: Risky change? Vulnerability and adaptation between climate change and transformation dynamics in Can Tho City, Vietnam, Stuttgart Germany, 2014.

Gilles, D., and Moore, M.: Review of Hydraulic Flood Modeling Software used in Belgium, The Netherlands, and The United Kingdom, International Perspectives in Water Resource Management, 15 [online] Available from: http://www.iihr.uiowa.edu/education1/international/UK/projects_files/ipwrsm_paper _gilles_moore_Dan_editted.pdf, 2010.

Giuliani. M., Herman. J.D., Quinn. J.D.: Kirsch-Nowak_Streamflow_Generator, https://github.com/julianneq/Kirsch-Nowak_Streamflow_Generator, 2017.

Gorgoglione, A., Gioia, A., Iacobellis, V., Piccinni, A.F., and Ranieri, E.: A rationale for pollutograph evaluation in ungauged areas, using daily rainfall patterns: Case Studies of the Apulian region in southern Italy. Appl. Environ. Soil Sci., 9327614, doi:10.1155/2016/9327614, 2016.

Grossi, P., Kunreuther, H., and Windeler, D.: An introduction to catastrophe models and insurance, in: Catastrophe modeling: a new approach to managing risk, edited by: Grossi, P. and Kun- reuther, H., Springer Science Business Media, Inc., Boston, 23–42, 2005.

Grünthal, G., Thieken, A.H., Schwarz, J., Radtke, K.S., Smolka, A., and Merz, B.: Comparative risk assessments for the city of Cologne - Storms, floods, earthquakes. Nat. Hazards, 38, 21–44, doi:10.1007/s11069-005-8598-0, 2006.

Gumbel, E.J.: Les valeurs extrêmes des distributions statistiques, Annales de l'Institut Henri Poincaré, 5 (2): 115–158, 1935.

Hall, J.W., Dawson, R.J., Sayers, P.B., Rosu, C., Chatterton, J.B., and Deakin, R.: A methodology for national-scale flood risk assessment, Water Maritime Eng., 156(3), 235–247, doi:10.1680/maen.156.3.235.37976, 2003.

Hallegatte, S., Green, C., Nicholls, R. J. and Corfee-Morlot, J.: Future flood losses in major coastal cities, Nat. Clim. Chang., 3(9), 802–806, doi:10.1038/nclimate1979, 2013.

Hapuarachchi, H.A.P., Takeuchi, K., Zhou, M.C., Kiem, A.S., Georgievski, M., Magome, J., and Ishidaira, H.: Investigation of the Mekong River basin hydrology for 1980–2000 using the YhyM, Hydrol. Process, 22, 1246–1256, 2008.

Hegger, D. L. T., Driessen, P. P. J., Wiering, M., Van Rijswick, H. F. M. W., Kundzewicz, Z. W., Matczak, P., Crabbé, A., Raadgever, G. T., Bakker, M. H. N., Priest, S. J., Larrue, C. and Ek, K.: Toward more flood resilience: Is a diversification of flood risk management strategies the way forward?, Ecology and Society, 21(4), doi:10.5751/ES-08854-210452, 2016.

Hirabayashi, Y., Mahendran, R., Koirala, S., Konoshima, L., Yamazaki, D., Watanabe, S., Kim, H., and Kanae, S.: Global flood risk under climate change, Nat. Clim. Chang., 3, 4–6, doi:10.1038/nclimate1911, 2013.

Hoang, L. P., Lauri, H., Kummu, M., Koponen, J., Vliet, M. T. H. Van, Supit, I., Leemans, R., Kabat, P., and Ludwig, F.: Mekong River flow and hydrological extremes under climate change, Hydrol. Earth Syst. Sci., 3027–3041, doi:10.5194/hess-20-3027-2016, 2016.

Hoang, L.P., Biesbroek, R., Tri, V.P.D., Kummu, M., van Vliet, M.T.H., Leemans, R., Kabat, P., and Ludwig, F.: Managing flood risks in the Mekong Delta: How to address emerging challenges under climate change and socioeconomic developments, Ambio., 1–15, doi:10.1007/s13280-017-1009-4, 2018.

Hoang, L. P., Vliet, M. T. H. Van, Kummu, M., Lauri, H., Koponen, J., Supit, I., Leemans, R., Kabat, P. and Ludwig, F.: Science of the Total Environment The Mekong ' s future fl ows under multiple drivers : How climate change , hydropower developments and irrigation expansions drive hydrological changes, Science of the Total Environment, 649, 601–609, doi:10.1016/j.scitotenv.2018.08.160, 2019.

Hoanh, C.T, Guttman. H, Droogers. P and Aerts. J.: Water, climate, food, and environment in the Mekong basin in Southeast Asia: contribution to the project ADAPT: adaptation strategies to changing environments. Final report, No H041917, IWMI Research Reports, International Water Management Institute, 2003.

Hoanh, C.T., Jirayoot, K., Lacomne, G., and Srunetr, V.: Impacts of Climate Change and Development on Mekong Flow Regimes First Assessment—2009; MRC Management Information Booklet Series No. 4; Mekong River Commission: Vientiane, Laos, 2010.

Hunter, N.M., Bates, P.D., Horritt, M.S., and Wilson, M.D.: Improved simulation of flood flows using storage cell models. Proceedings of the ICE – Water Management 159: 9–18, 2006.

Huong, H.T.L., and Pathirana, A.: Urbanization and climate change impacts on future urban flooding in Can Tho city, Vietnam, Hydrol. Earth Syst. Sci., 17, 379–394, doi:10.5194/hess-17-379-2013, 2013.

ICPR.: Atlas of flood danger and potential damage due to extreme floods of the Rhine. International Commission for the Protection of the Rhine (ICPR), Koblenz, 2001.

IPCC.: IPCC Special Report – Emission Scenarios. Summary for Policymakers. A Special Report of IPCC Working Group III Published for the Intergovernmental Panel on Climate Change (IPCC), 2000.

IPCC.: Working Group I Contribution to the IPCC Fifth Assessment Report, Climate Change 2013: The Physical Science Basis, Summary for Policymakers Geneva, Switzerland: IPCC Retrieved from http://www.climatechange2013.org/images/uploads/WGIAR5-SPM_Approved27Sep2013.pdf, 2013.

Javadnejad, F., Waldron, B., and Hill, A.: LITE Flood: Simple GIS-Based mapping approach for real-time redelineation of multifrequency floods, Nat Hazards Rev., 18(3), doi:10.1061/(ASCE)NH.1527-6996.0000238, 2017.

Johnson, J. M., Moore, L. J., Ells, K., Murray, A. B., Adams, P. N., Mackenzie, R. A. and Jaeger, J. M.: Recent shifts in coastline change and shoreline stabilization linked to storm climate change, Earth Surface Processes and Landforms, 40(5), 569–585, doi:10.1002/esp.3650, 2015.

Kaneko, S. and Toyota, T.: Long-term urbanization and land subsidence in Asian megacities: An indicators system approach, in: Groundwater and Subsurface Environments., Springer, Tokyo, doi:10.1007/978-4-431-53904-9_13, 2011.

Khue, N. N.: Modelling of tidal propagation and salinity intrusion in the Mekong main sstuarine system, Technical paper of Mekong Delta Sallinity Intruction Studies, Phase II, Mekong Secretariat, Bangkok, 51 pp, 1986.

Kind, J. M.: Economically efficient flood protection standards for the Netherlands, J. Flood Risk Manage., 7(2), 103–117, doi:10.1111/jfr3.12026, 2014.

King, A. D., Donat, M. G., Fischer, E. M., Hawkins, E., Alexander, L. V., Karoly, D. J., Dittus, A. J., Lewis, S. C. and Perkins, S. E.: The timing of anthropogenic emergence in simulated climate extremes, Environmental Research Letters, 10(9), doi:10.1088/1748-9326/10/9/094015, 2015.

Kingston, D.G., Thompson, J.R., and Kite, G.: Uncertainty in climate change projections of discharge for the Mekong River Basin, Hydrol. Earth Syst. Sci., 15, 1459–1471, doi:10.5194/hess-15-1459-2011, 2011.

Kirsch, B. R., Characklis, G. W., and Zeff, H. B.: Evaluating the impact of alternative hydro-climate scenarios on transfer agreements: Practical improvement for generating synthetic streamflows, J. Water Resour. Plan. Manag, 139(4), 396-406, 2013.

Koks, E. E., Jongman, B., Husby, T. G. and Botzen, W. J. W.: Combining hazard, exposure and social vulnerability to provide lessons for flood risk management, Environmental Science and Policy, 47, 42–52, doi:10.1016/j.envsci.2014.10.013, 2015.

Kreibich, H., Piroth, K., Seifert, I., Maiwald, H., Kunert, U., Schwarz, J., Merz, B. and Thieken, A. H.: Is flow velocity a significant parameter in flood damage modelling?, Nat. Hazards Earth Syst. Sci., 9, 1679–1692, doi:10.5194/nhess-9-1679-2009, 2009.

Kron, W., Steuer, M., Löw, P., and Wirtz, A.: How to deal properly with a natural catastrophe database – analysis of flood losses, Nat. Hazards Earth Syst. Sci., 12, 535–550, https://doi.org/10.5194/nhess-12-535-2012, 2012.

Kuenzer, C., Gue, H., Huth, J., Leinenkugel, P., Li, X., and Cech, S.: Flood mapping and flood dynamic of the Mekong Delta: ENVISAT ASAR-WSM base time series analyses, Remote Sens., 5, 687–715, doi:10.3390/rs5020687, 2013.

Kundzewicz, Z. W., Kanae, S., Seneviratne, S. I., Handmer, J., Nicholls, N., and Peduzzi, P.: Flood risk and climate change: global and regional perspectives, Hydrol. Sci. J., 59, 1–28, doi:10.1080/02626667.2013.857411, 2014.

Kvočka, D., Falconer, R.A., and Bray, M.: Flood hazard assessment for extreme flood events, Nat Hazards., 84, 1569–1599, doi: 10.1007/s11069-016-2501-z, 2016.

Lang, M., Ouarda, T.B.M.J. and Bobée, B.: Towards operational guidelines for over-threshold modeling, J. Hydrol., 225 (3–4), 103–117, doi: 10.1016/S0022-1694(99)00167-5, 1999.

Lasage, R., Veldkamp, T. I. E., de Moel, H., Van, T. C., Phi, H. L., Vellinga, P., and Aerts, J. C. J. H.: Assessment of the effectiveness of flood adaptation strategies for HCMC, Nat. Hazards Earth Syst. Sci., 14, 1441–1457, doi:10.5194/nhess-14-1441-2014, 2014.

Lauri, H., De Moel, H., Ward, P.J., Räsänen, T.A., Keskinen, M., and Kummu, M.: Future changes in Mekong River hydrology: Impact of climate change and reservoir operation on discharge, Hydrol. Earth Syst. Sci., 16, 4603–4619, doi:10.5194/hess-16-4603-2012, 2012.

Le, T.V.H., Haruyama, S., Nguyen, H.N., and Tran, T.C.: Study of 2001 Flood Using Numerical Model in the Mekong River Delta, Vietnam, In Proceedings of the International Symposium on Floods in Coastal Cities under Climate Change Conditions, Khlong Nueng, Thailand, pp. 65–73, 2005.

Le, T.V.H., Nguyen, H.N., Wolanski, E.J., Tran, T.C., and Haruyama, S.: The combined impact on the flooding in Vietnam's Mekong River delta of local man-made structures, sea level rise, and dams upstream in the river catchment. Estuarine Coastal. Estuar. Coast. Shelf Sci., 71, 110–116, doi:10.1016/j.ecss.2006.08.021, 2007

Leandro, J., and Martins, R.: A methodology for linking 2D overland flow models with the sewer network model SWMM 5.1 based on dynamic link libraries, Water Sci. Technol., 73(12), 3017–3026, doi:10.2166/wst.2016.171, 2016.

Leelawat, N., Pee, L., and Iijima, J.: Mobile Apps in Flood Disasters: What Information do Users Prefer?, International Conference on Mobile Business, Berlin, Germany, 15, 2013.

Leiserowitz, A.: Climate Change Risk Perception and Policy Preferences: The Role of Affect, Imagery, and Values, Clim. Change., 77(1), 45-72, doi:10.1007/s10584-006-9059-9, 2006.

Lenderink, G., and Meijgaard, E. V. A. N.: Increase in hourly precipitation extremes beyond expectations from temperature changes, Nat Geosci., 511–514, doi:10.1038/ngeo262, 2008.

Lendering, K. T., Sebastian, A., Jonkman, S. N. and Kok, M.: Framework for assessing the performance of flood adaptation innovations using a risk-based approach, J. Flood Risk Manag., 12, doi:10.1111/jfr3.12485, 2019.

Leskens, J. G., Brugnach, M. and Hoekstra, A. Y.: Application of an interactive water simulation model in urban water management: A case study in Amsterdam, Water Science and Technology, 70(11), 1729–1739, doi:10.2166/wst.2014.240, 2014.

Leskens, J. G., Kehl, C., Tutenel, T., Kol, T., Haan, G. de, Stelling, G., and Eisemann, E.: An interactive simulation and visualization tool for flood analysis usable for practitioners, Mitig Adapt Strateg Glob Chang., 22(2), 307–324, doi:10.1007/s11027-015-9651-2, 2017.

Löwe, R., Urich, C., Kulahci, M., Radhakrishnan, M., Deletic, A., and Arnbjerg-Nielsen, K.: Simulating flood risk under non-stationary climate and urban development conditions – Experimental setup for multiple hazards and a variety of scenarios, Environ. Model. Softw., 102, 155–171, doi:10.1016/j.envsoft.2018.01.008, 2018.

Maloney, E. K., Lapinski, M. K., and Witte, K.: Fear Appeals and Persuasion: A review and update of the extended parallel process model, Soc. Personal. Psychol. Compass, 5(4), 206-219, doi:10.1111/j.1751-9004.2011.00341.x, 2011.

McMillan, H.K., and Brasington J.: Reduced complexity strategies for modelling urban floodplain inundation, Geomorphology, 90, 226–243, 2007.

McNamara. D.E. and Keeler. A.: A coupled physical and economic model of the response of coastal real estate to climate risk, Nature Clim Change, 3(6), 559–562, 10.1038/nclimate1826, 2013

MDP.: Mekong Delta Plan: Long-term vision and strategy for a safe, prosperous and sustainable delta, Vietnam, 126 pp, 2013.

Melbourne Water.: Property flood level information: https://www.melbournewater.com.au/planning-and-building/apply-to-build-or-develop/property-flood-level-information, last access: 6 August 2020.

Menzel, L., Niehoff, D., Bürger, G., and Bronstert, A.: Climate change impacts on river flooding: a modelling study of three meso-scale catchments, in: Climatic Change: Implications for the Hydrological Cycle and for Water Management, edited by: Beniston, M., Springer, Netherlands, pp. 249–269, doi: 10.1007/0-306-47983-4_14, 2002.

Merkens, J., Reimann, L., Hinkel, J. and Vafeidis, A. T.: Gridded population projections for the coastal zone under the Shared Socioeconomic Pathways, Glob. Planet. Change, 145, 57–66, doi:10.1016/j.gloplacha.2016.08.009, 2016.

Merz, B., and Thieken, A.H.: Flood risk analysis: concepts and challenges, Österreichische Wasser-und Abfallwirtschaft., 56(3–4):27–34, 2004.

Merz, B., Thieken, A., and Gocht, M.: Flood risk mapping at the local scale: Concepts and Challenges, in: Flood Risk Management in Europe, Advances in Natural and Technological Hazards Research, vol 25, edited by: Begum, S., Stive, M.J.F., Hall, J.W., Springer, Dordrecht, 231–251, doi: 10.1007/978-1-4020-4200-3_13, 2007.

Merz, B., Hall, J., Disse, M., and Schumann, A.: Fluvial flood risk management in a changing world, Nat. Hazards Earth Syst. Sci., 10, 509–527, doi:10.5194/nhess-10-509-2010, 2010.

Meyer, V., Haase, D., and Scheuer, S.: Flood risk assessment in European river basins-- concept, methods, and challenges exemplified at the Mulde River, Integr Environ Assess Manag., 5(1), 17-26, doi:10.1897/ieam_2008-031.1, 2009.

Min, S-K., Zhang, X., Zwiers, F.W., and Hegerl, G.C.: Human contribution to more-intense precipitation extremes, Nature., 470(7334), 378-381, doi:10.1038/nature09763, 2011.

Minderhoud, P. S. J., Erkens, G., Pham, V. H., Vuong, B. T. and Stouthamer, E.: Assessing the potential of the multi-aquifer subsurface of the Mekong Delta (Vietnam) for land subsidence due to groundwater extraction, Proc. Int. Assoc. Hydrol. Sci., 372, 73–76, doi:10.5194/piahs-372-73-2015, 2015.

Mintzberg, H., Raisinghani, D. and Theoret, A.: The Structure of 'un-structured' Decision Processes, Administrative Science Quarterly, 21(2), 246–275, 1976.

Mishra, K., and Sinha, R.: Geomorphology Flood risk assessment in the Kosi megafan using multi-criteria decision analysis : A hydro-geomorphic approach, Geomorphology, 350, 106861, doi:10.1016/j.geomorph.2019.106861, 2020.

MONRE: Climate Change and Sea Level Rise Scenarios, Science, 294(5547), 1379–1388, 2016.

Moore, J. L.: Cost–Benefit Analysis: Issues in Its Use in Regulation, CRS Report 95–760, June, 1995.

Mora, C., Spirandelli, D., Franklin, E. C., Lynham, J., Kantar, M. B., Miles, W., Smith, C. Z., Freel, K., Moy, J., Louis, L. V, Barba, E. W., Bettinger, K., Frazier, A. G., Ix, J. F. C., Hanasaki, N., Hawkins, E., Hirabayashi, Y., Knorr, W., Little, C. M. and Emanuel, K.: Hazards intensified by greenhouse gas emissions, Nat. Clim. Chang., doi:10.1038/s41558-018-0315-6, 2018.

Mosavi, A., Ozturk, P., and Chau, K.-W.: Flood Prediction Using Machine Learning Models: Literature Review, Water., 10, doi:10.3390/w10111536, 2018.

MRC.: Overview of the Hydrology of the Mekong Basin, Mekong River Commission, Vientiane, Laos, 2005.

MRC.: State of the Basin Report 2010, Mekong River Commission, Vientiane, Laos, 2010a.

MRC.: Assessment of Basin-wide Development Scenarios, Basin Development Plan Programme Phase 2, Mekong River Commission, Vientiane, Laos, 2010b.

Muis, S., Verlaan, M,, Winsemius, H.C., Aerts, J. C. J. H., and Ward, P. J.: A global reanalysis of storm surges and extreme sea levels, Nat Commun., doi:10.1038/ncomms11969, 2016.

Nash, J.E.,and Sutcliffe, J.V.: River flow forecasting through conceptual models: Part 1. A discussion of principles. J. Hydrol., 10, 282–290, doi:10.1016/0022-1694(70)90255-6, 1970.

Neal, J., Villanueva, I., Wright, N., Willis, T., Fewtrell, T. and Bates, P.: How much physical complexity is needed to model flood inundation ?, Hydrol. Processes, 26, 2264–2282, doi:10.1002/hyp.8339, 2012.

Neumann, B., Vafeidis, A. T., Zimmermann, J. and Nicholls, R. J.: Future Coastal Population Growth and Exposure to Sea-Level Rise and Coastal Flooding - A Global Assessment, PLoS One., 10(6), doi:10.1371/journal.pone.0118571, 2015.

Ngo, H., Pathirana, A., Zevenbergen, C., Ranasinghe, R.: An Effective Modelling Approach to Support Probabilistic Flood Forecasting in Coastal Cities — Case Study: Can Tho, Mekong Delta, Vietnam, J. Mar. Sci. Eng., 1–19, doi:10.3390/jmse6020055, 2018.

Ngo, C. C., Poortvliet, P. M., and Feindt, P. H.: Drivers of flood and climate change risk perceptions and intention to adapt: an explorative survey in coastal and delta Vietnam, J. Risk Res., 23(4), 424-446. doi:10.1080/13669877.2019.1591484, 2020a.

Ngo, H., Pathirana, A., Ranasinghe, R., and Radhakrishnan, M.: Inform, an instant flood risk modelling tool for Can Tho city in Mekong Delta, Vietnam, http://fg.srv.pathirana.net/, last access: 5 August 2020b.

Nguyen, H.N.: Human Development Report 2007/2008 Flooding in Mekong River Delta, Viet Nam, Human Development Report, United Nations Development Programme: New York, NY, USA, 2008, Volume 4.

Nguyen, H.Q., Huynh, T.T.N., Pathirana, A., and Van der Steen, P.: Microbial risk assessment of tidal—induced urban flooding in can Tho city (Mekong delta, Vietnam). Int. J. Environ. Res. Public Health., 14, 1–10, doi:10.3390/ijerph14121485, 2017a.

Nguyen, H.Q., Radhakrishnan, M., Huynh, T.T.N., Baino-Salingay, M.L., Ho, L.P., Van der Steen, P., and Pathirana, A.: Water quality dynamics of urban water bodies during flooding in can Tho City, Vietnam, Water (Switzerland)., 9, 1–12, doi:10.3390/w9040260, 2017b.

Nguyen, H. Q., Radhakrishnan, M., Bui, T. K. N., Tran, D. D., Ho, L. P., Tong, V. T., . . . Ho, H. L.: Evaluation of retrofitting responses to urban flood risk in Ho Chi Minh City using the Motivation and Ability (MOTA) framework, Sustain. Cities Soc., 47, 101465, doi:10.1016/j.scs.2019.101465, 2019.

Nguyen, N.Q.: Planning, developing new urban areas in peri-urban areas and the problem of urban flooding - An Khanh and An Hoa case studies, Ninh Kieu district, Can Tho city, http://www.cantholib.org.vn:84/Ebook.aspx?p=27B9F975353796A6E64627B93B65 654746C6B65637B91B857557, 2016.

Nicholls, R.J.: Coastal Flooding and Wetland Loss in the 21st Century: Changes under the SRES Climate and Socio-Economic Scenarios, Global Environmental Change., 14, 69-86, doi: 10.1016/j.gloenvcha.2003.10.007, 2004.

Nicholls, R.J., Wong, P.P., Burkett, V.R., Codignotto, J.O., Hay, J.E., McLean, R.F., Ragoonaden, S., and Woodroffe, C.D.: Coastal systems and low-lying areas. In Climate Change 2007: Impacts, Adaptation and Vulnerability, Contribution of Working Group II to the Fourth Assessment Report of the Intergovernmental Panel on Climate Change, Cambridge University Press: Cambridge, UK, 2007.

Nicholls, R.J., Adger, W.N., Hutton, C.W., Hanson, S.E.: Delta Challenges and Trade-Offs from the Holocene to the Anthropocene, in: Deltas in the Anthropocene, edited by: Nicholls, R., Adger, W., Hutton, C., Hanson, S., Palgrave Macmillan, Cham, UK, doi:10.1007/978-3-030-23517-8_1, 2020.

Nied, M., Schröter, K., Lüdtke, S., Nguyen, V. D. and Merz, B.: What are the hydro-meteorological controls on flood characteristics?, Journal of Hydrology, 545, 310–326, doi:10.1016/j.jhydrol.2016.12.003, 2017.

Nisbet, M. C.: Communicating climate change: Why frames matter for public engagement, Environ Sci Policy, 51(2), 12-23, doi:10.3200/ENVT.51.2.12-23, 2009.

NIURP.: Development Strategies (CDS) for Medium-Size Cities in Vietnam: Can Tho and Ha Long, Vietnam Ministry of Construction, Hanoi, Vietnam, 2010.

Nowak, K., Prairie, J., Rajagopalan, B. and Lall, U.: A nonparametric stochastic approach for multisite disaggregation of annual to daily streamflow, Water Resour. Res., 46(8), doi:10.1029/2009WR008530, 2010.

OECD.: Financial Management of Flood Risk, OECD Publishing, Paris, doi:10.1787/9789264257689-en, 2016.

Panagoulia, D. and Dimou, G.: Sensitivity of flood events to global climate change, Journal of Hydrology, 191(1–4), 208–222, doi:10.1016/S0022-1694(96)03056-9, 1997.

Pappenberger, F., Dutra, E., Wetterhall, F. and Cloke, H. L.: Deriving global flood hazard maps of fluvial floods through a physical model cascade, Nat Hazards Earth Syst Sci., 4143–4156, doi:10.5194/hess-16-4143-2012, 2012.

Pasquier, U., He, Y., Hooton, S., Goulden, M. and Hiscock, K. M.: An integrated 1D–2D hydraulic modelling approach to assess the sensitivity of a coastal region to compound flooding hazard under climate change, Nat. Hazards, 98(3), 915–937, doi:10.1007/s11069-018-3462-1, 2019.

Pathirana, A.: SWMM5-EA-A tool for learning optimization of urban drainage and sewerage systems with genetic algorithms, in: Proceedings of the 11th International Conference on Hydroinformatics, New York, NY, USA, 17–21 August 2014, CUNY Academic Works: New York, NY, USA, 2014.

Pathirana, A., Denekew, H.B., Veerbeek, W., Zevenbergen, C., and Banda, A.T.: Impact of urban growth-driven landuse change on microclimate and extreme precipitation – A sensitivity study, Atmos. Res., 138, 59–72, doi:10.1016/j.atmosres.2013.10.005, 2014.

Penning-Rowsell, E. C., Fordham, M., Correia, F. N., Gardiner, J., Green, C., Hubert, G., Ketteridge, A.-M., Klaus, J., Parker, D. Peerbolte, B., Pflugner, W., Reitano, B., Rocha, J., Sanchez- Ar-cilla, A., Saraiva, M. d. G., Schmidtke, R., Torterotot, J.- P., van der Veen, A., Wierstra, E., and Wind, H.: Flood hazard assessment, modelling and management: Results from the EUROflood project, in Floods across Europe: Flood hazard assessment, modelling and management, edited by: Penning-Rowsell, E. C., and Fordham, M., Middlesex University Press, London, 1994.

Penning-Rowsell, E., Floyd, P., Ramsbottom, D., and Surendran, S.: Estimating injury and loss of life in floods: a deterministic framework, Nat Hazards., 36:43–64, doi: 10.1007/s11069-004-4538-7, 2005.

Piman, T., Lennaerts, T., and Southalack, P.: Assessment of hydrological changes in the lower Mekong basin from basin-wide development scenarios, Hydrol. Process., 27, 2115–2125, doi:10.1002/hyp.9764, 2013.

Pokhrel, Y., Burbano, M., Roush, J., Kang, H., Sridhar, V., and Hyndman, D.: A review of the integrated effects of changing climate, land use, and dams on Mekong river hydrology, Water., 10, 266, doi:10.3390/w10030266, 2018.

Prudhomme, C., Crooks, S. and Kay, A. L.: Climate change and river flooding: Part 1 Classifying the sensitivity of British catchments Climate change and river flooding: part 1 classifying the sensitivity of British catchments, Clim. Change., doi:10.1007/s10584-013-0748-x, 2013.

Radhakrishnan, M., Pathirana, A., Ashley, R., and Zevenbergen, C.: Structuring climate adaptation through multiple perspectives: Framework and Case study on flood risk management, Water., 9(2), 129, Retrieved from http://www.mdpi.com/2073-4441/9/2/129, 2017.

Radhakrishnan, M., Nguyen, H. Q., Gersonius, B., Pathirana, A., Vinh, K. Q., Ashley, R. M., and Zevenbergen, C.: Coping capacities for improving adaptation pathways for flood protection in Can Tho, Vietnam, Clim Change., doi:10.1007/s10584-017-1999-8, 2018a.

Radhakrishnan, M., Islam, T., Ashley, R. M., Pathirana, A., Quan, N. H., Gersonius, B., and Zevenbergen, C.: Context specific adaptation grammars for climate adaptation in urban areas, Environ Modell Softw., 102, 73-83, doi:10.1016/j.envsoft.2017.12.016, 2018b.

Ramsbottom, D., Floyd, P., and Penning-Rowsell, E.: Flood risks to people: Phase 1. R&D Technical Report FD2317, Department for the Environment, Food and Rural Affairs (DEFRA), UK Environment Agency, 2003.

Ranasinghe, R.: Assessing climate change impacts on open sandy coasts: A review. Earth Sci. Rev., 160, 320–332, doi:10.1016/j.earscirev.2016.07.011, 2016.

Ranasinghe, R. and Jongejan, R.: Climate Change, Coasts and Coastal Risk, J. Mar. Sci. Eng., 6–9, doi:10.3390/jmse6040141, 2018.

Ranasinghe, R., Wu, C. S., Conallin, J., Duong, T. M. and Anthony, E. J.: Disentangling the relative impacts of climate change and human activities on fluvial sediment supply to the coast by the world's large rivers: Pearl River Basin, China, Scientific Reports, 9(1), 1–10, doi:10.1038/s41598-019-45442-2, 2019.

Räsänen, T.A., Koponen, J., Lauri, H., and Kummu, M.: Downstream hydrological impacts of hydropower development in the Upper Mekong Basin, Water Resour. Manag., 26, 3495–3513, doi:10.1007/s11269-012-0087-0 2012.

Räsänen, T.A., Someth, P., Lauri, H., Koponen, J., Sarkkula, J., and Kummu, M.: Observed river discharge changes due to hydropower operations in the Upper Mekong Basin. J. Hydrol., 545, 28–41, doi:10.1016/j.jhydrol.2016.12.023, 2017.

Rijcken, T., Stijnen, J. and Slootjes, N.: "Simdelta"-Inquiry into an internet-based interactive model for water infrastructure development in the Netherlands, Water (Switzerland), 4(2), 295–320, doi:10.3390/w4020295, 2012.

Rijcken, T.: EMERGO: the Dutch flood risk system since 1986, Ph.D. thesis, Delft University of Technology, the Netherlands, 188 pp., 2017.

Rijke, J., van Herk, S., Zevenbergen, C., Ashley, R., Hertogh, M. and ten Heuvelhof, E.: Adaptive programme management through a balanced performance/strategy oriented focus, International Journal of Project Management., 32(7), 1197–1209, doi:10.1016/j.ijproman.2014.01.003, 2014.

Rijkswaterstaat.: Over overstroomik?: https://www.overstroomik.nl/, last access: 6 August 2020.

Risk Management Solutions and Lloyd's.: Coastal Communities and Climate Change: Maintaining Future Insurability, Lloyd's, London., 28, 2008.

Rodela, R., Ligtenberg, A. and Bosma, R.: Conceptualizing serious games as a learning-based intervention in the context of natural resources and environmental governance, Water (Switzerland), 11(2), doi:10.3390/w11020245, 2019.

Rojas, R., Scientific, T. C. and Watkiss, P.: Climate change and river floods in the European Union : Socio-economic consequences and the costs and benefits of adaptation, Glob Environ Change, 1737–1751, doi:10.1016/j.gloenvcha.2013.08.006, 2013.

Rossman, L.A.: Storm Water Management Model User's Manual, EPA: Washington, DC, USA, 353 pp, 2015.

Samuels, P. P.: Stakeholder involvement in flood risk management – contribution from the FLOODsite project, (April), 1–13, 2012.

SCE Can Tho (Vietnam).: Comprehensive Resilience Planning For Integrated Flood Risk Mangement - Final Report WorldBank, 2013.

SCE.: Comprehensive Resilience Planning For Integrated Flood Risk Mangement for Can Tho, https://sce.fr/en/comprehensive-resilience-planning-integrated-flood-risk-management-can-tho-vietnam, last access: 7 August 2020.

Scheffran, J.: Tools for stakeholder assessment and interaction, Stakeholder Dialogues in Natural Resources Management., 153–185, doi:10.1007/978-3-540-36917-2_6, 2007.

Seijger, C., Hoang, V. T. M., van Halsema, G., Douven, W., and Wyatt, A.: Do strategic delta plans get implemented? The case of the Mekong Delta Plan, Reg. Environ. Change., 19(4), 1131-1145, doi:10.1007/s10113-019-01464-0, 2019.

Seyoum, S. D., Vojinovic, Z., Price, R. K., and Weesakul, S.: Coupled 1D and noninertia 2D flood inundation model for simulation ofurban flooding, Journal of Hydraulic Engineering, vol. 138, no. 1, pp. 23–34, 2012.

Small, C., and Nicholls, R.J.: A global analysis of human settlement in coastal zones. J. Coast. Res., 19, 584–599, 2003.

Spence, A., Poortinga, W., and Pidgeon, N.: The Psychological Distance of Climate Change. Risk Analysis, 32(6), 957-972, doi:10.1111/j.1539-6924.2011.01695.x, 2012.

Stern, N.: The economics of climate change: The stern review, The Economics of Climate Change: The Stern Review, 9780521877251, 1–692, doi:10.1017/CBO9780511817434, 2007.

Suarez, P., Ribot, J.C. and Patt, A.G.: Climate information, equity and vulnerability reduction, in: Ruth. M and Ibarrarian, M.E., eds. Distribution impacts of climate change and disasters: concepts and cases. Northampton: Edward Elgar, 151–165. United, 2009.

Suriya, S., Mudgal, B.V.: Impact of Urbanization on Flooding: The Thirusoolam Sub Watershed – A Case Study, J. Hydrol., 412–413, 210–219, doi:10.1016/j.jhydrol.2011.05.008, 2012.

Sutherland, J., Walstra, D.J.R., Chesher, T.J., van Rijn, L.C., and Southgate, H.N.: Evaluation of coastal area modelling systems at an estuary mouth, Coast. Eng., 51, 119–142, doi:10.1016/j.coastaleng.2003.12.003, 2004.

Syvitski, J.P.M., Kettner, A.J., Overeem, I., Hutton, E.W.H., Hannon, M.T., Brakenridge, G.R., Day, J., Vörösmarty, C., Saito, Y., Giosan, L., and Nicholls, R.J.: Sinking deltas due to human activities, Nature Geosci., 2, 681-686, doi: 10.1038/ngoe629, 2009.

Takagi, H., Ty, T. V, Thao, N. D. and Esteban, M.: Ocean tides and the influence of sea-level rise on floods in urban areas of the Mekong Delta, J. Flood Risk Manage., 8, 292–300, doi:10.1111/jfr3.12094, 2015.

Thaler, T. and Levin-Keitel, M.: Multi-level stakeholder engagement in flood risk management-A question of roles and power: Lessons from England, Environ Sci Policy, 55, 292–301, doi:10.1016/j.envsci.2015.04.007, 2016.

Tiggeloven, T., Moel, H. De, Winsemius, H.C., Eilander, D., Erkens, G., Gebremedhin, E., Loaiza, A.D., Kuzma, S., Luo, T., and Bouwman, A.: Global-scale benefit – cost analysis of coastal flood adaptation to different flood risk drivers using structural measures, Nat. Hazards Earth Syst. Sci., 20, 1025–1044, doi:10.5194/nhess-20-1025-2020, 2020.

Timmerman, J. G., Beinat, E., Termeer, C. J. A. M. and Cofino, W. P.: A methodology to bridge the water information gap, Water Science and Technology, 62(10), 2419–2426, doi:10.2166/wst.2010.513, 2010.

Toombes, L. and Chanson, H.: Numerical Limitations of Hydraulic Models, in: The 34th International Association for Hydraulic Research World Congress, Brisbane, Australia, 2011.

Tran Anh, D., Van, S. P., Dang, T. D., and Hoang, L. P.: Downscaling rainfall using deep learning long short-term memory and feedforward neural network, Int J Climatol., 39(10), 4170-4188, doi:10.1002/joc.6066, 2019.

Tran, P., Marincioni, F., Shaw, R., Sarti, M., and Van An, L.: Flood risk management in Central Viet Nam: Challenges and potentials, Nat. Hazards., 46, 119–138, doi:10.1007/s11069-007-9186-2, 2008.

Tri, V.P.D., Trung, N.H., and Tuu, N.T.: Flow dynamics in the Long Xuyen Quadrangle under the impacts of full-dyke systems and sea level rise, VNU J. Sci. Earth Sci., 28, 205–214, 2012.

UK Environment Agency.: Check you long term flood risk: https://flood-warning-information.service.gov.uk/long-term-flood-risk/, last access: 6 Aug 2020.

UNISDR.: 2013 floods a "turning point": https://undrr.org/news/2013-floods-turning-point, last access: 3 Aug 2020.

Valiela, I.: Global Coastal Change, Blackwell, Oxford, UK, 368 pp, 2006.

van den Belt, M.: Mediated modeling: a system dynamics approach to environmental consensus building, Washington, DC: Island Press, 2004.

van Berchum, E., van Ledden, M., Timmermans, J., Kwakkel, J., and Jonkman, S.: Rapid flood risk screening model for compound flood events in Beira, Mozambique, Nat. Hazards Earth Syst. Sci., 1–18, doi:10.5194/nhess-2020-56, 2020.

Van, P.D.T., Popescu, I., Van Griensven, A., Solomatine, D.P., Trung, N.H., and Green, A.: A study of the climate change impacts on fluvial flood propagation in the Vietnamese Mekong Delta, Hydrol. Earth Syst. Sci., 16, 4637–4649, 2012.

Västilä, K., Kummu, M., Sangmanee, C., and Chinvanno, S.: Modelling climate change impacts on the flood pulse in the lower Mekong floodplains, J. Water Clim. Chang., 1, 67–86, doi:10.2166/wcc.2010.008, 2010.

Vorogushyn, S., Bates, P. D., de Bruijn, K., Castellarin, A., Kreibich, H., Priest, S., Schröter, K., Bagli, S., Blöschl, G., Domeneghetti, A., Gouldby, B., Klijn, F., Lammersen, R., Neal, J. C., Ridder, N., Terink, W., Viavattene, C., Viglione, A., Zanardo, S. and Merz, B.: Evolutionary leap in large-scale flood risk assessment needed, Wiley Interdisciplinary Reviews: Water, 5(2), e1266, doi:10.1002/wat2.1266, 2018.

Vuik, V., Jonkman, S. N., Borsje, B. W., and Suzuki, T.: Nature- based flood protection: the efficiency of vegetated foreshores for reducing wave loads on coastal dikes, Coast. Eng., 116, 42–56, doi:10.1016/j.coastaleng.2016.06.001, 2016.

Ward, P.J., De Moel, H., and Aerts, J.C.J.H.: How are flood risk estimates affected by the choice of return-periods?, Nat. Hazards Earth Syst. Sc., 11, 3181–3195, doi.org/10.5194/nhess-11-3181-2011, 2011a.

Ward, P. J., Marfai, M. A., Yulianto, F., Hizbaron, D. R., and Aerts, J. C. J. H.: Coastal inundation and damage exposure estimation: a case study for Jakarta, Nat. Hazards, 56, 899–916, doi:10.1007/s11069-010-9599-1, 2011b.

Ward, P.J., Jongman, B., Sperna Weiland, F.C., Bouwman, A., Van Beek, R., Bierkens, M., Ligtvoet, W., and Winsemius, H.C.: Assessing flood risk at the global scale: model setup, results, and sensitivity, Environ Res Lett., 8(4), doi:10.1088/1748-9326/8/4/044019, 2013.

Wassmann, R., Hien, N.X., Hoanh, C.T., and Tuong, T.P.: Sea level rise affecting the Vietnamese Mekong Delta: Water elevation in the flood season and implications for rice production, Clim. Chang., 66(1-2), 89–107, doi:10.1023/B:CLIM.0000043144.69736.b7, 2004.

Wind, H. G., Nieron, T. M., De Blois, C. J., and De Kok, J. L.: Analysis of flood damages from the 1993 and 1995 Meuse floods, Water Resour. Res., 35, 3459–3465, doi:10.1029/1999WR900192, 1999.

Wong, P.P., Losada, I.J., Gattuso, J.P., Hinkel, J., Khattabi, A., McInnes, K.L., Saito, Y., Sallenger, A.: Coastal Systems and Low-Lying Areas. In Climate Change 2014: Impacts, Adaptation, and Vulnerability. Part A: Global and Sectoral Aspects, Contribution of Working Group II to the Fifth Assessment Report of the Intergovernmental Panel on Climate Change, Cambridge University Press: Cambridge, UK, New York, NY, USA, 2014.

Zope, P.E., Eldho, T.I., and Jothiprakash, V.: Impacts of land use-land cover change and urbanization on flooding: A case study of Oshiwara River Basin in Mumbai, India. Catena, 145, 142–154, doi:10.1016/j.catena.2016.06.009, 2016.

List of acronyms

1D	One Dimensional
2D	Two Dimensional
EADs	Expected Annual Damages
FRM	Flood Risk Management
FRRS	Flood risk reduction strategies
Inform	The instant Flood risk modelling tool
IPCC	The Intergovernmental Panel on Climate Change
MDP	Mekong Delta Plan
MONRE	Vietnamese Ministry of Natural Resources and Environment
MRC	Mekong River Commission
LECZs	Low elevation coastal zones
NHMS	National Hydro-meteorological Service of Viet Nam
NIURP	National Institute for Urban and Rural Planning
SWMM	EPA Storm Water Management Model
VRSAP	Vietnam River System and Plains

List of Tables

153

List of Figures

About the author

Hieu Quang Ngo was born in Ha Noi, Viet Nam. He obtained his Bachelor's degree and MSs' degree in Hydraulic Engineering from Water Resources University, Viet Nam. From 2007 to 2015, he worked at Hydraulic Construction Institute – Vietnam Academy for Water Resources. In this period, he functioned as an engineering designer, researcher and project leader, where he was involved in designing hydraulic engineering works and doing scientific research related to planning and designing hydraulic structures. In 2015, he was offered a PhD position in IHE Delft, in collaboration with Delft University of Technology. His research interests include: climate change impacts and human activities on flooding in coastal and estuaries cities; development of efficient modelling systems to support constructing probabilistic flood hazard maps and, further, to support flood risk reduction management in coastal and estuaries cities; development of interactive tools for co-designing of risk reduction measures; remote sensing and GIS applications in flood management; coastal flood modelling at global, regional and local scales.

Journals publications

Ngo, H., Pathirana, A., Zevenbergen, C., and Ranasinghe, R.: An Effective Modelling Approach to Support Probabilistic Flood Forecasting in Coastal Cities - Case Study: Can Tho, Mekong Delta, Vietnam. J. Mar. Sci. Eng, 6, 55, 2018.

Journals sumitted

Ngo, H., Ranasinghe, R., Zevenbergen, C., Kirezci, E., Maheng, D., Radhakrishnan, M., and Pathirana, A.: An efficient modelling approach for probabilistic assessments of present-day and future fluvial flooding, Frontiers in climate, 2021.

Ngo, H., Radhakrisnan, M., Ranasinghe, R., Pathirana, A., Zevenbergen, C.: Instant Flood Risk Modelling (*Inform*) Tool for Co-design of Flood Risk Management Strategies with Stakeholders, Water, 2021

Conference proceedings

Ngo, H., Bomers, A., Augustijn, D.C.M., Ranasinghe, R., Filatova, T., and Hulscher, S.J.M.H.: Reconstruction of the 1374 Rhine river flood event using a 1D-2D hydraulic modelling. Proceedings of 8th International Conference on Flood Management 2021, Online, USA, 10 August 2021.

Ngo, H., Bomers, A., Augustijn, D.C.M., Ranasinghe, R., Filatova, T., and Hulscher, S.J.M.H.: Improving historic flood reconstruction using a detailed 1D-2D coupled hydraulic model approach. Proceedings of the International Conference on the Status and Future of the World's Large Rivers 2021, Online, Moscow, Russia, 3-6 August 2021.

Ngo, H., Bomers, A., Augustijn, D.C.M., Ranasinghe, R., Hulscher, S.J.M.H., and Filatova, T.: The effect of climate change on flood risk perception: a case study of the Grensmaas using a coupled hydraulic agent-based model. Proceedings of the Joint International Resilience Conference 2020, Online, the Netherlands, 23–27 November 2020.

Ngo, H., Pathirana. A., Zevenbergen, C., and Ranasinghe, R.: Probabilistic flood risk maps under climate change scenarios – A modelling study of Can Tho City, Viet Nam. Proceedings of the Water and Development Congress & Exhibition 2019, Colombo, Sri Lanka, 01–05 December 2019.

Huynh, N., Nguyen, Q., Vinh, V., **Ngo, H.**, Baker, S., and Pathirana, A.: Microbial pollution in flood-related waters in urban areas – A case study in Ninh Kieu district, Can Tho City of Viet Nam. Proceedings of the Water and Development Congress & Exhibition 2019, Colombo, Sri Lanka, 01–05 December 2019.

Ranasinghe, R., Bosboom, J., Uhlenbrook, S., Roelvink, D., **Ngo, H.**, and Stive, M. J. F.: A scale aggregated model to estimate climate change driven coastline change along inlet interrupted coasts. Proceedings of Coastal Sediments 2011, Miami, FL, USA, pp. 286-298, 2011.

Netherlands Research School for the
Socio-Economic and Natural Sciences of the Environment

D I P L O M A

for specialised PhD training

The Netherlands research school for the
Socio-Economic and Natural Sciences of the Environment
(SENSE) declares that

Ngo Quang Hieu

born on 5th June 1983 in Ha Noi, Vietnam

has successfully fulfilled all requirements of the
educational PhD programme of SENSE.

Delft, 22 November 2021

Chair of the SENSE board

Prof. dr. Martin Wassen

The SENSE Director

Prof. Philipp Pattberg

The SENSE Research School has been accredited by the Royal Netherlands Academy of Arts and Sciences (KNAW)

K O N I N K L I J K E   N E D E R L A N D S E
A K A D E M I E   V A N   W E T E N S C H A P P E N

The SENSE Research School declares that Ngo Quang Hieu has successfully fulfilled all requirements of the educational PhD programme of SENSE with a work load of 46.3 EC, including the following activities:

SENSE PhD Courses

- Environmental research in context (2019)
- Research in context activity: 'Development of an interactive, web-based flood risk management tool for co-designing with stakeholders' (2021)

Selection of Other PhD and Advanced MSc Courses

- PhD Start-up Module A, B and C,TU Delft (2016-2018)
- Water Sensitive Cities, IHE Delft (2018)
- English Writing Course, IHE Delft (2018)
- Advanced CHI professional training workshop on PCSWMM and EPA SWMM5 modeling, Computational Hydraulics International (2018)
- Designing Scientific Posters and lay-out for Theses with Adobe InDesign, TU Delft (2019)
- The Informed Researcher: Information and Data Skills, TU Delft (2019)
- The Art of Presenting Science, TU Delft (2019)
- Cross Cultural Communication skills in Academia, TU Delft (2019)
- Online Self-Presentation: Focus, Structure and divers Visualisations, TU Delft (2020)
- Popular Scientific Writing, TU Delft (2020)
- Decision Making Under Uncertainty: Introduction to Structured Expert Judgment, EDX (2020)
- Flood Risk Management, IHE Delft (2020)

Management and Didactic Skills Training

Coaching MSc students with the group work in the MSc program (2018-2020)

Selection of Oral Presentations

- *An Integrated 1D/2D modelling to support constructing probabilistic extreme flood hazard maps for Can Tho City for 2050.* PhD symposium, 10-11 October 2019, Delft, The Netherlands
- *Probabilistic flood risk maps under climate change scenarios - A modelling study of Can Tho City, Vietnam.* IWA Water and Development Congress & Exhibition, 01-05 December 2019, Colombo, Sri Lanka
- *The effect of climate change on flood risk perception: a case study of the Grensmaas using a coupled hydraulic & agent-based model.* Joint international Resilience Conference, 23-27 November 2020, Online, The Netherlands

SENSE coordinator PhD education

Dr. ir. Peter J. Vermeulen